Power

Seven internationally renowned writers address the theme of Power from the perspective of their own disciplines. Energy expert Mary Archer begins with an exploration of the power sources of our future. Astronomer Neil Tyson leads a tour of the orders of magnitude in the cosmos. Mathematician and inventor of the Game of Life John Conway demonstrates the power of simple ideas in mathematics. Screenwriter Maureen Thomas explains the mechanisms of narrative power in the media of film and videogames, Elisabeth Bronfen the emotional power carried by representations of life and death, and Derek Scott the power of patriotic music and the mysterious Mozart effect. Finally, celebrated parliamentarian Tony Benn critically assesses the reality of power and democracy in society.

THE DARWIN COLLEGE LECTURES

Power

Edited by *Alan Blackwell* and *David MacKay*

CAMBRIDGE
UNIVERSITY PRESS

CAMBRIDGE UNIVERSITY PRESS

Cambridge, New York, Melbourne, Madrid, Cape Town, Singapore, São Paulo

Cambridge University Press
The Edinburgh Building, Cambridge CB2 2RU, UK

Published in the United States of America by Cambridge University Press, New York

www.cambridge.org
Information on this title: www.cambridge.org/9780521823777

First published 2005

Printed in the United Kingdom at the University Press, Cambridge

A catalogue record for this publication is available from the British Library

ISBN-13   978-0-521-82377-7 hardback
ISBN-10   0-521-82377-3 hardback

Contents

Introduction

ALAN F. BLACKWELL AND DAVID J. C. MacKAY

Space, time and power are fundamentals of physics that determine the dynamic structure of our lives. Recent publications from the Darwin College Lecture Series have addressed two of these topics: *Space* in the 2001 lecture series and *Time* in 2000. Each of those volumes included a range of perspectives that span the arts, humanities and sciences. Now, in this new volume, we have invited seven international authorities to analyse and interpret the theme of *Power* as it is understood in their different fields of learning. The subjects that they consider include not only the sources of power that humanity has at its disposal, but also the forms of power that are exerted over us by cultural products and societies.

Life on earth, and of course all human activity, depends on the availability of sufficient power to support that activity. Mary Archer starts our exploration of power by considering where this power comes from. Drawing both on her academic work as a researcher in chemistry and Professor of Energy Policy, and on her public life including presidency of the National Energy Foundation, Archer reviews and forecasts human power usage and supply. Her chapter on the future of sustainable power sources addresses the rate with which we consume fossil fuel resources, and the alternatives that might supply the hundreds of exajoules we consume each year.

Grand topics of discussion require big numbers. An exajoule is 10 to the power of 18 – 1 000 000 000 000 000 000 or a billion billion – joules. In order to relate human experience to global phenomena, let alone the scale of our primary energy sources in the sun and other stars, we must use the language of mathematics. Neil deGrasse Tyson, director of the Hayden Planetarium in New York, gives us a tour of the power offered by this language, expressed

Power, edited by Alan Blackwell and David MacKay. Published by Cambridge University Press.
© Darwin College 2005.

1

as powers of ten, that extends our understanding of scale to the largest and smallest limits of the universe. John Conway, Professor of Mathematics at Princeton University, investigates the descriptive power of mathematics itself, beyond the structure of numbers and measurement, to the power that mathematical reasoning gives us to define, describe and predict the structures of our life.

The abstract mathematical world is becoming more real to us through the developments of digital technology. Digital technologies not only describe, but enhance and transform, our experience of the physical world, creating new universes of the imagination. Maureen Thomas, in the fourth chapter of this book, analyses the narrative power behind these universes. She surveys the history of narrative through the technologies that create new human experiences, from the two-dimensional screen of the moving image, to the development of three-dimensional camera motions that draw the viewer into and beyond the screen, and finally the fourth dimension of narrative control in interactive videogames. Elisabeth Bronfen then looks beyond narrative, to the whole range of literary devices that structure our aesthethic experience. Indeed Bronfen's chapter approaches the absolute limit of experience – the final and incommunicable experience of death. As she explains, the representation of death in art offers us a degree of power over its personal and social consequences.

Both Bronfen and Thomas draw our attention, through the insights of critical analysis, to the power that can be extended over an audience by artistic techniques. This power influences us all, whether or not we are aware of it. Even the abstract art of music, without any direct power of representation, becomes a channel for the expression of power. Music, however, is associated not only with power over our thoughts and emotions, but over whole societies. As Derek Scott demonstrates in his chapter on the power of music, even this abstract and non-representational art becomes a powerful medium for moral, behavioural and political persuasion.

Ultimately, the greatest concentration of power over our lives is the power exerted by our fellow men and women, through national and, increasingly, global politics. Tony Benn, one of our most experienced and respected parliamentarians, writes of the origins of power in society, the proper foundations of power for all peoples in the world and, as appropriate in a closing chapter, the opportunities we have to exert some power over our own futures.

It has been our privilege to work with these international leaders and thinkers, and we are grateful to Darwin College for having given us this opportunity. We are also grateful to Sally Thomas, Joseph Bottrill and Vincent Higgs at Cambridge University Press for their help in editing this collection from the Darwin Lecture Series, and to Themis Halvantzi for her valuable assistance with picture research.

1 Sustainable power

MARY ARCHER

Energy is vital to our economic and social well-being. Economic growth would be impossible without the ready availability of fuels to provide affordable heat, light and motive and electrical power. Yet the provision of power from fossil fuels poses a major threat to our environment, for we live, most of us now accept, in a globally warming world. Low-carbon technology will be essential in powering tomorrow's world in a sustainable way.

It has been said that extrapolation of the present North American per capita energy consumption to the world's population of 6 billion people would require the resources of several additional Earths to cope with the waste products, in particular the CO_2 (carbon dioxide) emissions that come from burning fossil fuels. Clearly that is unsustainable. Sustainable development, a concept popularised by the Brundlandt Report of 1987, is generally understood to mean development that enables us to meet our present needs without compromising the ability of future generations to meet their own needs. The UK government has been committed to sustainable development since 1994, and now monitors national progress in achieving sustainability in energy supply and consumption, as well as in other areas of life. 'National emissions of greenhouse gases' is one of the Department of Trade and Industry's (DTI's) six headline indicators of sustainability.

But would a sustainable energy policy pay for the sequestration of anthropogenic CO_2 emissions or for raising coastal defences to cope with rising sea levels? Would it encourage nuclear power on the grounds that it is CO_2-free or terminate it on account of the hazardous wastes it produces? Would it have us shivering in our thermal underwear in the dim light of one energy-efficient light bulb, or would it have us streaking around in hydrogen-fuelled cars in a

Power, edited by Alan Blackwell and David MacKay. Published by Cambridge University Press.
© Darwin College 2005.

world of abundant clean energy from solar or wind power? One's answer to these questions depends on the view one takes of the damage that global warming will do, the prospects of discovering significant new reserves of fossil fuels, the future pace of innovation in energy technology and technology transfer, the commercial prospects for renewables, and the political and economic risks involved in an expansion of nuclear power. In my view, sustainability in energy supply will best be achieved by maximising the future role of renewable energy, and thus I shall concentrate on technical innovation and the renewables in this chapter.

Current world energy consumption

To demonstrate the ways in which our current energy supply is unsustainable, I shall borrow an idea from the late Carl Sagan's book *The Dragons of Eden* and compress the long history of the world into one calendar year. On this time scale, 1 January represents the formation of the Earth by the condensation of interstellar matter some four and a half billion years ago, and it is now midnight on New Year's Eve. Life appeared on 9 February, 40 days after the formation of the Earth, and primitive photosynthetic organisms on 1 March. The Carboniferous Period, during which the Earth's deposits of coal, oil and gas were laid down by the decomposition of vegetable and animal matter, spanned the first fortnight of December.

On this time scale, quite a lot happens on New Year's Eve, as it tends to do in real life. *Homo sapiens sapiens*, presumably accompanied by *Femina sapienta sapienta*, emerged out of Africa 23 minutes ago. The Industrial Revolution, which marks the beginning of man's exploitation of the Earth's fossil fuel deposits, began just 1.5 seconds ago. Within the next few seconds – within the next few centuries in real time – most of the rest of this pre-packed solar energy will inevitably be consumed. We are midnight's children, striking one short hydrocarbon match in the middle of a long, dark night. When that hydrocarbon match burns low, assuming we have not by then discovered some new exploitable force of nature, we will have to turn to some combination of nuclear and renewable energy.

We may reach the same conclusion from Figure 1.1, which depicts the world's energy resources on a logarithmic scale. On the left are shown the energy contents of the world's proven reserves of oil, gas, coal and fissile uranium

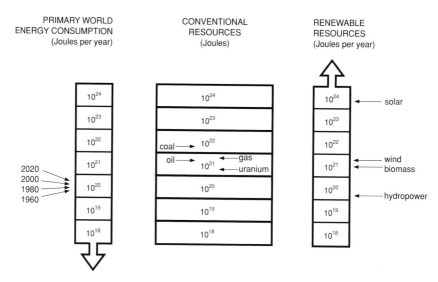

FIGURE 1.1 Global proven conventional resources of coal, oil, natural gas and uranium (used in non-breeder mode), as compared with world primary energy consumption, and the total energy per year available in the renewable resources of solar, wind, biomass and hydropower.

(that is, uranium used in thermal nuclear reactors, which unlike fast reactors do not 'breed' additional nuclear fuel). Also shown on the left are the world's primary energy consumption in 1960, 1980 and 2000, and predicted global consumption in 2020. It is clear that proven fossil fuel resources are within sight of being used up; at the current rate of usage, we have enough conventional oil and gas to last only a few more decades. More may well be discovered, and additional resources such as tar sands and methane hydrates may prove to be recoverable in an economically and environmentally acceptable way, but even then the hydrocarbon flame must go out within a few hundred years.

On the right of Figure 1.1 is the Earth's energy income from the renewables, that is, those energy sources that are powered by natural flows of energy from the sun and in the wind, waves and so on. At the top is the amount of solar energy absorbed by the Earth in one year, showing that the sun is by far our most abundant renewable energy resource. Every 44 minutes, sufficient energy from the sun strikes the Earth to provide the entire world's energy requirements for one year. The energy driving most of the other renewables, notably biomass, also derives partly or wholly from the sun.

Technical innovation in energy supply

In a globally warming world grown accustomed to cheap, subsidised fossil fuel power, incentives to encourage the use of renewable energy and more energy-efficient technology in a low-carbon economy are both necessary and justified. Internationally, the UK is signed up to the UN's Kyoto Protocol to curb the country's carbon dioxide emissions. Internally, the government has set the ambitious target of reducing national CO_2 emissions by 21.5 per cent by the year 2010 from the 1990 baseline. We have the (imperfect) beginnings of a carbon tax in the Climate Change Levy, and the new Renewable Energy Obligation will oblige electricity suppliers progressively to increase the proportion of their electricity sourced from renewables up to 10 per cent by 2010. Moving much beyond that figure would require our inflexible electrical transmission system to be reconfigured to accommodate fluctuating power inputs from renewables.

New methods of storing electricity would also be very helpful, and in this context the Cambridge region is host to a pioneering facility under construction at Little Barford, on the Cambridgeshire/Bedfordshire border. This is Regenesys, the world's largest secondary battery, with a charge/discharge cycle involving concentrated solutions of sodium polybromide/bromide and sodium sulphide/polysulphide, designed to store 120 MWh of electricity and discharge it at a nominal power rating of 15 MW.

On the horizon is the emergent hydrogen economy, brought closer by significant advances in the technology of PEMFCs (Proton Exchange Membrane Fuel Cells), likely to find their way into a new generation of electric cars over the next few years. Cambridge University itself will soon be entering the hydrogen era through its USHER project, in which photovoltaic modules placed on the roof of an architectural colonnade in the new West Cambridge campus will generate electrical power to electrolyse water and make hydrogen that will power buses to shuttle us to and from the new site.

Technical innovation in fossil fuel generation can also help to bring a lower-carbon energy economy closer. Thanks to continuing improvements in turbine materials and design, combined cycle gas power stations now generate electricity at nearly 60 per cent efficiency. Various 'clean coal' technologies are also under development. In one notable pilot scheme, coal in the North Dakota coalfields is gasified *in situ* by steam injection, and the resulting syngas ($CO + H_2$) is brought to the surface. The carbon monoxide (CO) in the syngas is further reacted with water to produce carbon dioxide (CO_2) and more hydrogen (H_2).

The hydrogen is separated and used as a fuel while the CO_2 is sent to the Weyburn oil field in Canada, where it is injected, at the same time enhancing oil recovery and sequestering the carbon.

Many commentators believe that the deep cuts in CO_2 emissions that may prove necessary in coming decades cannot be achieved without building new nuclear plant to replace the world's ageing stock of pressurised-water and gas-cooled reactors. A new generation of smaller, passively safe nuclear reactors such as the new CANDU reactor and the South African pebble-bed reactor are likely to be built and tested, and more radical innovations in nuclear power, such as moving from the uranium cycle to the thorium cycle, are possible. One day we may master fusion power, and experiments on plasma containment continue at JET, the Joint European Torus facility at Culham, and elsewhere.

The sun and thermal uses of solar energy

Meanwhile we already have a fully functional fusion reactor, our sun. In astronomical terms, the sun is no more than a middle-sized, middle-aged star, but to us it is overwhelmingly important. Like all main-sequence stars, the sun burns hydrogen to helium, destroying some mass in the process and creating energy in accordance with Einstein's seminal equation $E = mc^2$, where E is energy, m is mass and c is the speed of light. The sun is losing mass at the brisk rate of 4.5 billion kg per second and correspondingly beaming out about 10^{26} W of radiant power more or less uniformly into all directions in space. The Earth intercepts about one-billionth of this power, and it is this absorbed solar energy that keeps the Earth's surface at a habitable global mean temperature of 288 K instead of the 2.3 K characteristic of deep space and has enabled the creation and evolution of life on Earth.

Modern attempts to turn the sun's radiant energy into other energy forms date back to the nineteenth century. Several early collector designs, such as the Mouchot Sun Machine shown in Figure 1.2, were based on parabolic dishes aimed at the sun to concentrate its rays at a focal point, thereby reaching a temperature capable of driving an engine. In a modern solar power tower system, of which there are a few prototypes such as Eurelios, in Adrano, Sicily, and Solar Two in Barstow, California, a field of heliostats (mirrors) tracks the sun and reflects direct sunlight onto a receiver mounted on a central tower.

FIGURE 1.2 The Sun Machine invented by the French mathematician Augustin Mouchot, exhibited at the World Fair in Paris in 1878, could power a half-horsepower engine.

Power tower systems usually achieve solar concentration ratios of 300 to 1500, and operate at temperatures from 550 °C to 1500 °C, easily capable of driving a steam generator.

Focussing sunlight about one axis rather than two requires paraboloidal troughs rather than dishes. The Kramer Junction Company in the Mojave Desert in California is the world's largest solar thermal power station, turning out 354 MW electricity from east–west oriented troughs, along the focal line of which there runs a pipe containing a heat-transfer oil. This provides heat to a conventional steam generator, and back-up gas is used to provide constant output power on dull days.

Nearer home, a company in Foxton called HelioDynamics (www.hdsolar.com) is developing a range of hybrid solar thermal/electric converters, aiming for the commercial and industrial market in areas where stand-alone heat and power is needed because of poor security of electricity supply. The power comes from photovoltaic cells mounted at the focal spot of the solar concentrator, and the heat from cooling them.

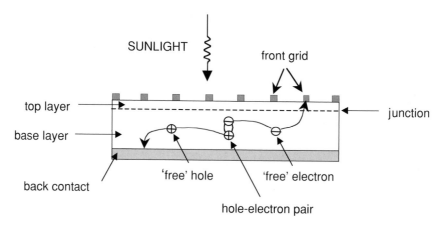

FIGURE 1.3 In a 'classical' solar cell, the junction between the electrically dissimilar top and base layers contains a built-in electric field that separates the hole–electron pairs generated by photon absorption.

Finally, as regards thermal applications of solar energy, flat-plate solar thermal collectors are widely used in sunny parts of the world to heat water for domestic and commercial use. In the UK, these are more cost-effective if self-built.

Direct conversion of solar energy: photovoltaic cells

Rather than using sunlight as a source of heat, one can use it as a stream of photons (quantised packets of radiant energy) in so-called direct conversion or photoconversion devices. These work not because they are heated by the sun, but because they separate hole–electron pairs generated by the absorption of solar photons, as shown schematically in Figure 1.3.

Photovoltaic cells, also known as solar cells, are by far the most successful man-made direct solar converters. These are flat, thin semiconductor devices, usually made of crystalline or amorphous silicon, that generate direct low-voltage electric power when illuminated. Hole–electron pair separation normally occurs at the junction between the top-layer and base-layer semiconductor, as shown in Figure 1.3. Photovoltaic power can be generated anywhere – in temperate or tropical locations, in urban or rural environments, in distributed or grid-feeding mode – and the modules are silent and non-polluting in operation. As a fuel-free distributed source of electricity, photovoltaic arrays

FIGURE 1.4 BP Harmony petrol stations, with a photovoltaic roof that provides power to the pumps and forecourt. Photograph: courtesy of Ben Hill, BP Solar.

could in the long run make a major contribution to carbon dioxide abatement. In the UK, for example, each kW installed avoids the emission of about 1 tonne of carbon dioxide per year.

Photovoltaic arrays have the additional advantage of being uniquely scalable, the only energy source that can supply power on a scale of milliwatts to MW from an easily replicated modular technology with excellent economies of scale in manufacture. Figure 1.4 shows a BP Harmony petrol station, where the pumps and forecourt are powered by the photovoltaic roof.

Photovoltaic cells were developed in the 1950s for use in space, and they are now widely used to provide on-board power to satellites. In the developing world and other areas without access to electrical supply, photovoltaic power can confer great benefits. Photovoltaic modules are also used in a range of professional applications to provide stand-alone power to remote locations. In the fast-growing BIPV (building-integrated photovoltaics) sector, photovoltaic

11

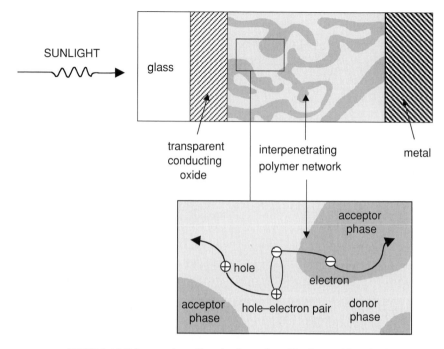

FIGURE 1.5 Polymer solar cell made of two ultra-thin sheets of doped polymer, in which hole–electron pair separation occurs at the extended phase boundary between the co-blended polymers. From J. J. M. Halls and R. H. Friend, 'Organic photovoltaic devices', in M. D. Archer and R. Hill, eds., *Clean Electricity from Photovoltaics* (Imperial College Press, 2001).

modules are mounted on the surface of built structures, rather than as free-standing arrays, thus avoiding the cost of conventional cladding.

Costs of photovoltaic technology have fallen steadily over recent decades, but the still-high cost remains the main barrier to wider deployment. Cumulative worldwide installed capacity is only just over 1 GW (providing less power than Sizewell B). Using thin films of multicrystalline semiconductors such as cadmium telluride to make solar cells instead of single-crystal silicon is one way forward. In the longer term, as Mr Maguire said to Ben Braddock in *The Graduate*, there's a great future in plastics. Organic polymers can now be made with adequate purity to be fabricated into ultra-thin co-blend solar cells in which hole–electron pairs are separated at the extended phase boundary between two different polymers, as shown in Figure 1.5.

Photosynthesis and biomass

Green leaves are nature's direct converters of solar energy. They take in atmospheric carbon dioxide and water vapour and, under the driving force of sunlight absorbed by the green pigment chlorophyll, photosynthesise these simple molecules into carbohydrates and other organic molecules. Animals eat green plants, or other animals that have eaten green plants, and 'burn' the stored carbohydrate in the process of metabolism, giving them the energy they need to stay alive.

Biomass is organic material, derived from plant or animal life, and is in effect chemically stored solar energy. It includes wood and wood waste, arable crops and grasses and animal manure, all of which if sufficiently dry can be burned as fuels. Biomass plays an important role in the world's energy economy, currently supplying about 14 per cent of global final energy consumption, and about 25 per cent in developing countries.

Wood

Wood is the largest store and source of biomass energy, and indeed of renewable energy, providing more than twice the contribution of hydroelectricity worldwide. Wood has been burned as a fuel since ancient times. In his book *A Forest Journey*, John Pelin charts the rise and fall of civilisations according to their access to timber. For 5000 years, forests fuelled the rise of civilisations as surely as deforestation signalled their decline. When the Romans ran out of silver to pay their legions in Spain and had to resort to a barter economy, it was not because the silver ore was exhausted, but because the smelters had deforested an area of 7000 square miles around them.

Fuelwood is often gathered and burned unsustainably but trees such as willow and poplar can be grown as sustainable crops in so-called Short Rotation Coppice (SRC) plantations. These trees grow so rapidly and readily from the 'stool' (base) that they can be harvested every three or so years and the chipped wood burned as a sustainable fuel. Sweden leads the European Union in the use of wood as a fuel, with some 18 000 hectares under SRC. The UK has about 2000 hectares under willow and poplar, mostly in the ARBRE project around Eggborough, which will provide 8 MW of electric power when it is fully operational.

Forests and their associated soils store a large quantity of carbon, and it has been suggested that we could capture the excess carbon dioxide we emit to the atmosphere by burning fossil fuels simply by growing more trees. Indeed, thousands of years of deforestation were finally halted in the 1990s, with the establishment of about 3 million hectares of new plantation forests. The Intergovernmental Panel on Climate Change (IPCC) has estimated that nearly 40 gigatonnes of carbon could be stored in new forest plantations over the next 50 years, but the mass of carbon in the atmosphere is increasing by about 3.5 gigatonnes per year, so planting trees could only store about 10 years' worth of man-made carbon emissions. A more immediate gain, in terms of abating these emissions, would be achieved by avoiding further deforestation and using reduced-impact logging techniques.

Energy crops

Energy crops are crops grown for their energy content rather than for food, or crop arisings used as a source of energy. Since the CO_2 produced when energy crops are burned is taken up by the growth of the next crop, they do not produce net CO_2 emissions, so burning them instead of fossil fuels would reduce CO_2 emissions in direct proportion to the quantity of fossil fuels replaced. Several countries with excess agricultural land have introduced incentives for growing energy crops on set-aside land.

In the UK, we boast the world's largest straw-burning power station, the Elean Power Station opened in early 2002. This will burn over 200 000 tonnes of straw per year obtained from local farmers (this represents only 2 per cent of the UK's surplus straw) and generate 36 MW of electricity, enough to power the city of Cambridge twice over.

Figure 1.6 shows the promising energy crop *Miscanthus*, sometimes known as elephant grass, a fast-growing perennial grass that can be grown in poor soil and a wide range of climatic conditions. This has been grown as an experimental crop at ADAS Arthur Rickwood in Mepal near Ely for ten years, with encouraging results. Its energy ratio, the ratio of energy produced to energy consumed in cultivation, turns out to be excellent at 25 : 1 (much better than SRC at 10 : 1). Soon the Elean Power Station will become the world's first *Miscanthus*-fed power station, supported by a 500-hectare plot grown by local farmers.

Biodiesel

Finally, as regards plant biofuels, there is biodiesel, which can be made from any oilseed crop. The oil from such plants is usually consumed as a food but

FIGURE 1.6 Experimental plot of the rapidly growing energy crop *Miscanthus*.
Photograph courtesy of Dr Mike Bullard, ADAS Arthur Rickwood.

with processing it can also be used as a fuel. Indeed, when Rudolph Diesel
designed his prototype diesel engine nearly a century ago, he intended it to
run on peanut oil. The era of petroleum-based diesel intervened, and it is only
recently that interest in biodiesel has revived. The EU produces about 900 000
tonnes of biodiesel per year, mainly in Austria. In the UK, a subsidy of about
20p per litre would be needed to make biodiesel competitive with conventional
diesel. Because oilseed crops are annual and the oil comes from the seeds only,
the energy ratio of biodiesel is poor at only 2:1. Genetic modification holds
out the possibility of increasing the seed yield, and hence improving the energy
ratio, of these crops.

Animal biomass

Large quantities of biomass wastes are available from farmed animals such as
chickens. Chickens are extremely efficient food processors: for every 2 lb of feed

they consume, they put on 1 lb of meat, making their so-called feed conversion ratio 1 : 2. Turkeys are nearly as efficient, at 1 : 2.6. Pigs come in at 1 : 4 and cows, which are very inefficient because they have to process grass through several stomachs, trail badly at 1 : 8.

The animal feed that does not get turned into meat is of course excreted, and impressive quantities of animal biomass are thereby produced. East Anglia alone produces 500 million broiler chickens a year, and they in turn produce 800 000 to 900 000 tonnes of litter. Spreading all this on the land could lead to high levels of nitrates and phosphates in the ground water. However, poultry litter has an energy content similar to that of wood and can be burned. Thetford is the home of the UK's largest chicken-litter-fuelled power station, Fibrothetford, which single-handedly generates about 8 per cent of the UK's renewable electricity by burning chicken litter and other wastes.

There are two problems with waste-to-energy schemes. The first is conflicting regulation. The UK's Renewable Energy Obligation permits purely organic waste to be converted to energy, but the EU's new Waste Incineration Directive rules that converting waste to energy by incineration will not be allowed. This may make the combustion of biomass unacceptable for no good scientific reason.

The second problem is one that often afflicts renewable energy – cost. Biomass-to-energy schemes look expensive because other fuels are or have been so heavily subsidised, and they are indeed expensive because there is little experience in manufacturing the combustion plant and poor infrastructure for collecting the biomass. The capital cost of Fibrothetford was £1250 per kW, compared with a modern coal-powered power station at about £800 per kW.

Atmosphere and ocean

Wind power

Winds are second-hand solar energy, arising from the non-uniform heating of the Earth by the sun combined with the dynamic effect of the Earth's rotation. They blow most strongly at high latitudes and altitudes and in marine locations. Historically, the extraction of mechanical power from wind goes back many centuries. Modern wind turbines, such as the 1.5 MW Ecotricity turbine at Swaffham, Norfolk, shown in Figure 1.7, produce electrical rather than mechanical power, by means of a generator sited in the chamber at the top

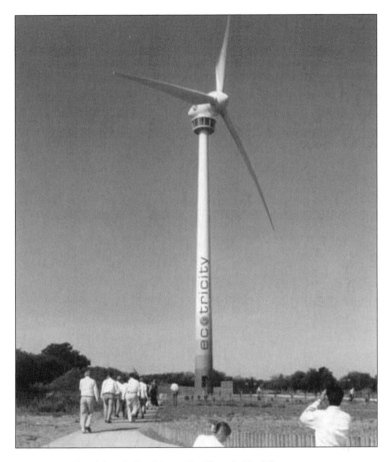

FIGURE 1.7 Ecotricity wind turbine at Swaffham in Norfolk.

of the tower. On-shore wind is the fastest-growing source of electrical power in the world, with about 13 500 MW installed worldwide. Twelve per cent of Danish electrical supply now comes from wind and this is planned to rise to 40 per cent by 2030.

On-shore wind is one of the cheapest renewable electricity resources; larger installations in favourable locations produce electrical power at little more than 2 pence per kWh, but on-shore wind projects have struggled to get planning permission in the UK, largely on grounds of visual intrusiveness. The next phase of development is likely to take place off-shore, where the Crown Estates have granted leases to a number of wind developers.

Wind turbines work at their maximum efficiency at one particular wind speed, called the rated speed, but because the wind does not always blow at the rated speed, on time average wind turbines deliver only about one-third of their rated power. (Advocates of nuclear power tend to disparage wind power on these grounds, although roughly two-thirds of the energy in the uranium consumed in nuclear power stations is discharged as waste heat, and only about one-third is turned into electricity.) Another charge sometimes brought against wind is the noise made by the turning blades, but a visit to a modern wind farm will soon disabuse you of that notion.

Marine tidal currents

The power in flowing rivers has been used to drive water turbines from medieval times, and it is now possible to adapt wind turbine technology to generate electricity from underwater currents in rivers and the oceans. Underwater turbines would have some advantages over wind turbines: they would deliver predictable power, they would not have to withstand sudden gusts, and they would be invisible from the shore. Also, water is nearly 850 times as dense as air, giving a 3-knot current about the same kinetic energy as a 100-mile-an-hour wind. A tidal current turbine would therefore be much smaller than a wind turbine with the same power rating (Figure 1.8 shows the comparative sizes) and the costs should be highly competitive with off-shore wind.

The best place to put an underwater turbine is between a rock and a hard place, where the tidal flow is strong. There are a number of good sites off the west of Britain, and the DTI is funding the construction near Lynmouth of a 300 kW prototype underwater turbine developed by Peter Fraenkel's company Marine Current Turbines.

Wave power

If wind power is second-hand solar energy, wave power is third-hand, because the long rolling waves of the deep ocean are created by the winds. Waves carry an impressive amount of energy, some 40–70 kW per metre, and the British Isles have excellent wave energy potential since they lie at the end of long fetches of the Atlantic Ocean. To extract energy from the waves, there must be relative movement of one part of the conversion device with respect to the other by means of some kind of hinge or flexible spine, and the natural frequency of the device should match the frequency at which waves arrive. Achieving this

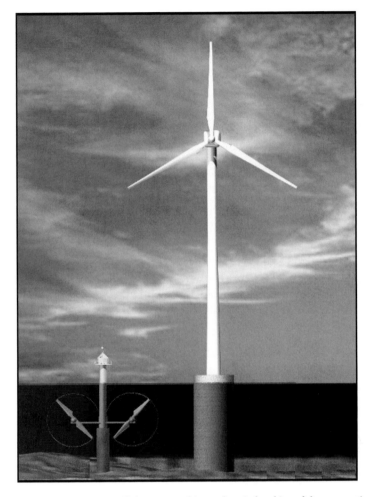

FIGURE 1.8 A marine tidal current turbine and a wind turbine of the same rating shown to scale. Diagram: courtesy of Dr Peter Fraenkel, Marine Current Turbines.

has proved challenging, but there are now a number of designs alongside the classic Salter Duck developed by Professor Stephen Salter of the University of Edinburgh in the 1970s.

The island of Islay has the only commercial wave power station in the world, Wavegen's 500 kW Limpet, based on a Wells Turbine, which has the unique property of turning in the same direction whichever way flow passes over the blades. A cluster of these stations is to be developed off the Western Isles.

In conclusion

The world needs power, and yet it cannot prudently allow the untrammelled growth of the use of fossil fuels. A new generation of nuclear plants is possible, but may not be politically or economically feasible. Meanwhile, almost all our renewable energy resources, with the exception of tidal and geothermal energy, derive directly or indirectly from the sun. When we have consumed pre-packaged solar energy – fossil fuels – and maybe before, if the worst fears of the climate modellers are realised, we will have to turn to some combination of nuclear and renewable energy. Former President Gerald Ford allegedly once remarked, "Solar energy is not going to happen overnight." He is right in more ways than one, but nonetheless, when we do adopt a sustainable power policy, the sun and its renewable offspring can rise to the challenge.

FURTHER READING

Houghton, J. T. *et al.* (2001). *Climate Change 2001: The Scientific Basis.* Cambridge: Cambridge University Press.

Performance and Innovation Unit (2002). *Energy Review* (www.piu.gov.uk/2002/energy/report).

United Nations Development Programme (UNDP) *World Energy Assessment: Energy and the Challenge of Sustainability* (www.undp.org/seed/eap/activities/wea/index.html).

2 Powers of ten

NEIL DeGRASSE TYSON

A few years ago, astronomers and astrophysicists did not agree on the age of
the universe. Some said it might be 10 billion years old, others said 20 billion.
You might think that we were completely clueless, not to know by a factor of
two how old the universe is. But you have to consider that no-one was arguing
the universe might be a trillion years old, or a quadrillion years, or a hundred
years old. We were only within a factor of two of each other, and this was
a pretty good thing. We knew we were nearing agreement. In fact the most
recent data indicate an age of 14 billion years, plus or minus one or two. In the
universe, quantities of time, size, temperature and distance come in such a vast
range that factors of two between friends are not important.

Introducing powers of ten

In this chapter, we're going to cover that whole vast range. But if we're going to
get through the entire universe in a few pages, factors of ten are the smallest
differences we should worry about.

$$10^0 = 1$$

We'll start here, the number 1. This needs no introduction. The number 1 has
no zeros to follow it, so we can write it as ten to the zeroth power. That zero
tells us how many zeros follow the 1, if you're going to write it out. This fact
turns out to be very important later on. Now let's go up by powers of a thousand.

$$10^3 = 1000$$

One with three zeros is one thousand. We have the international system of
prefixes – *kilo*. And of course we use this regularly: kilometre, one thousand

Power, edited by Alan Blackwell and David MacKay. Published by Cambridge University Press.
© Darwin College 2005.

metres, that's fine. This number needs no introduction, because it exists in our everyday lives.

$$10^6 = 1\,000\,000$$

Add another three zeros, to get ten to the power of six, a million. *Mega*. The population of New York city is about 8 million. London might be 10 million. We've seen these numbers before. Being a millionaire is not what it used to be, I'm told, by those who are. But it's certainly better than being a thousandaire.

$$10^9 = 1\,000\,000\,000$$

Let's go up another three zeros. Nine zeros, ten to the power of nine, *giga*, a billion. Billion is an important number in astronomy, because it shows up everywhere. There are a hundred billion stars in the galaxy. There are a hundred billion galaxies. We hear about billions on Earth from time to time: the net worth of Bill Gates is in the hundred billion category. He hasn't quite reached a hundred billion, but he'll get there, if he lives a natural life.

$$10^{12} = 1\,000\,000\,000\,000$$

In fact, Bill Gates will become the world's first trillionaire. Even if he sold all his Microsoft stock and bought conservative savings bonds, he would still be a trillionaire before he dies. A trillion has another three zeros: ten to the power of twelve, or *tera*. In the year that you turn 31 years old, you will live your billionth second. But you can't count to a trillion. It would take you 31 000 years. A trillion is about how many seconds have elapsed since cavemen roamed the earth.

$$10^{15} = 1\,000\,000\,000\,000\,000$$

Quadrillion. Working our way up the international system of prefixes: *peta*. Ten to the fifteenth power. You can calculate how many words could come out of someone's mouth. We often say that, in politics, a lot of words come out of a lot of people's mouths. Now if you add up all the sounds and words uttered by all the human beings that ever lived since the dawn of the human species, you'd get a hundred quadrillion. And this even includes what goes on in Parliament.

$$10^{18} = 1\,000\,000\,000\,000\,000\,000$$

My favourite number is this: quintillion, with the prefix *exa*. This figure is the estimated number of grains of sand on an average beach. You might ask how I know? There are ways to approximate this. You count how many grains of sand are in a cubic centimetre, then you estimate how many cubic centimetres are on the beach. We do this back of the envelope calculation (or back of the bathing-suit calculation) all the time in the sciences.

$$10^{21} = 1\,000\,000\,000\,000\,000\,000\,000$$

Sextillion. This is the estimated number of stars in the universe. It outstrips the number of grains of sand on the beach, and the number of sounds and words ever uttered by human beings. The scale of the universe is enormous. This number makes a powerful argument if you're debating the existence of life elsewhere in the universe. If you are so egocentric as to presume that life on Earth is the only life in the universe, just spend a night alone with this number.

$$10^{-3} = 0.001$$

You can of course go in the opposite direction, calculating fractions by powers of ten. Decimal places get shifted to the left now. One-tenth, 0.1, has its decimal point shifted one place to the left, so is written 10^{-1}. The standard prefix *deci* comes from the same origin as the decimal point – they both describe tenths. 10^{-2}, 0.01 or one-hundredth, is shifted another place to the left – *centi*. A centimetre is one hundredth of a metre. An American coin is called the cent because there are a hundred to the dollar. And one more decimal place, 10^{-3}, *milli*, is one-thousandth. The millimetre is small, but it's still useful in everyday life.

$$10^{-6} = 0.000\,001$$

One-millionth. *Micro* – they invented the microscope to see really small things. You'd think they could have invented the megascope to see big things. That would be a much more impressive name than telescope.

$$10^{-9} = 0.000\,000\,001$$

Here's a word that we're hearing a lot these days. *Nano*. One-billionth. Nano-technologies are the attempt to create tools that will let us manipulate molecules. A nanosecond – a billionth of a second – is the time that it takes light to travel 1 foot. That's a good way to remember the speed of light: 1 foot per nanosecond.

$$10^{-12} = 0.000\,000\,000\,001$$

Let's keep going down. One-trillionth, *pico*. Do you know that there is nothing in the universe that measures one-trillionth of a metre in size? There's a gap in how big things are. A picometre is smaller than an atom, and bigger than an atom's nucleus. I've encountered this gap in my work, as you'll see later in this chapter.

$$10^{-15} = 0.000\,000\,000\,000\,001$$

Ten to the power of minus 15 is about the size of the proton. We're getting down to fundamental particles now. *Femto*. A femtometre, ten to the minus fifteen metres, is one-quadrillionth of a metre.

$$10^{-18} = 0.000\,000\,000\,000\,000\,001$$

Ten to the power of minus 18. There is nothing measured that's this small, although there are things that we are pretty sure are smaller than this, like electrons, or quarks. We have yet to measure the dimensions of an electron. We know their behaviour, we know where they've been, we know where they're going. You'll never see one, but they're there. In modern science, were left behind the idea that seeing is believing. Evidence is not seeing, evidence is measuring – your retina is irrelevant to science.

$$10^{-21} = 0.000\,000\,000\,000\,000\,000\,001$$

One-sextillionth. I have no idea what in the physical universe this could represent, but we do have a standard prefix for it – *zepto*.

Table 2.1 recaps the international standard prefixes. We've got 48 powers of ten here. The measured sizes of all things that exist in the universe extend over 40 powers of ten. So everything in the universe fits within the range of names that we already have, and there are some to spare. That's kind of fascinating, or perhaps wishful thinking, that we might go on to measure things big enough or small enough that we need to use those extra words.

$$10^{81}$$

If you take all the atoms in a star, and multiply that number by all the stars in a galaxy, and then multiply that number by all the galaxies in the universe, you get this number. This is the total number of atoms there are, plus or minus a power of ten. The total number of atoms in the universe. Could you possibly need a number bigger than this? What would you be counting that wouldn't be contained within this number? As far as I know, this number hasn't been named, although I might vote for "totillion."

Table 2.1. *Forty-eight powers of ten.*

yocto	y	10^{-24}
zepto	z	10^{-21}
atto	a	10^{-18}
femto	f	10^{-15}
pico	p	10^{-12}
nano	n	10^{-9}
micro	μ	10^{-6}
milli	m	10^{-3}
centi	c	10^{-2}
deci	d	10^{-1}
deka	da	10^{1}
hecto	h	10^{2}
kilo	k	10^{3}
mega	M	10^{6}
giga	G	10^{9}
tera	T	10^{12}
peta	P	10^{15}
exa	E	10^{18}
zetta	Z	10^{21}
yotta	Y	10^{24}

10^{100}

This next number does have a name. Ten to the hundredth power. *Googol* is the official name of this number. It dwarfs the number of atoms in the universe, by 19 further powers of ten. So why would anyone need it? Well, it's just a fun number, and it's got a fun name. By the way, the search engine on the Internet, they have misspelled it 'G-o-o-g-l-e'. 'Googol' is the correct spelling.

$10^{10^{100}}$

Numbers can get still higher. This is one of the biggest numbers ever named, *googolplex*. Googol is ten to the hundredth power – that was 1 followed by a hundred zeros. But a googolplex is ten to the power of a googol. That's a one with a googol zeros. A googol zeros is more zeros than there are atoms in the universe. Nobody could ever write this number out . . . where would they get the ink?

You know how you can get even bigger numbers in nature? You don't keep counting objects, you count events. For example, if you're playing chess, how many possible chess games can there be? Far, far more than the number of chess pieces. So you're not counting things, you're counting configurations or events.

$$10^{10^{10^{34}}}$$

This gives us a number that dwarfs the googolplex! Skewes' number. Ten to the 10 to the 10 to the 34th power. And why do we have a number this size? If you imagine the universe, with all of its atoms, as a cosmic chessboard, and you ask the question how many possible combinations of configurations of atoms exist in the universe, you get this number. So in a way, this number is a measure of the total information content of the visible universe, because information relates to configuration of the states of matter. Either in your brain, in a computer, in a beaker, or in the whole universe.

Comprehending powers of ten

Can we use these numbers to help people appreciate the universe? There is a famous educational film, ten minutes long, called *Powers of Ten*. It's a zoom out from the Earth to the edge of the universe, and then back down into an atom, which is sitting in someone's hand on the beach. That video has a precedent that goes back to 1915. Henry Norris Russell, head of the department of astrophysics at Princeton, wrote a letter to the head of the American Museum of Natural History, where I work:

> Professor Osborne, your friendly interest in some of the ideas I spoke of the other day leads me to send you a sketch of my idea for a series of models or diagrams, of progressively smaller scales, to illustrate astronomical distances and the like. The enclosed scheme is entirely tentative, but might serve as a basis for consideration. It suggests the construction of a set of diagrams and models, each one one-hundredth the scale of the last.

$$10^2 \text{ to } 10^{10}$$

In his letter, he goes on to cover some of the ground that we have reviewed. At 100, 10^2, he suggests a plan of the hall in which the exhibit is situated. The hall is a hundred times bigger than the plan, so you get to compare the two. 10^4– a

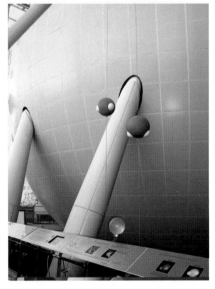

FIGURE 2.1 Size scales in the Hayden Planetarium, image courtesy of Hayden planetarium, Neil deGrasse Tyson.

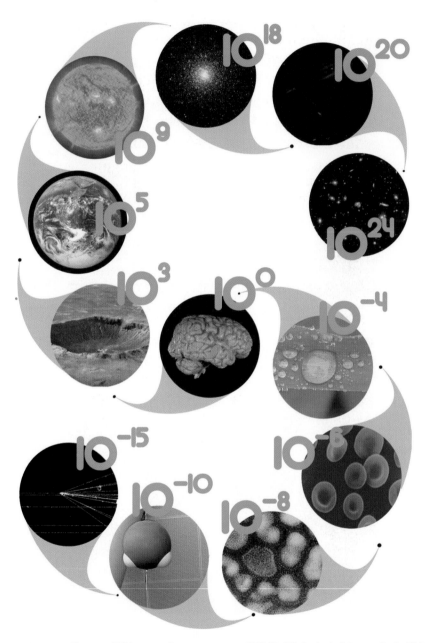

FIGURE 2.2 Powers of 10 in space. Images courtesy of NASA, US Geological Survey (D. Roddy), National Institutes of Health, pdphoto.org, Drs. Noguchi, Rodgers, and Schechter of NIDDK, CDC, National Center for Infectious Diseases (Dr. Erskine Palmer), Hayden Planetarium (Neil deGrasse Tyson), CERN.

FIGURE 2.3 Ascending from Earth. Continued over page. Images courtesy of NASA.

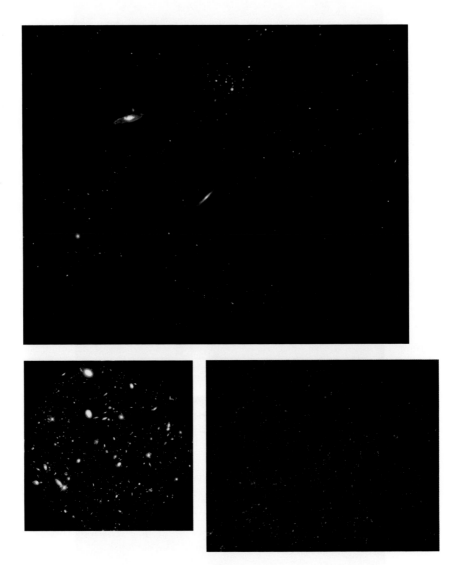

FIGURE 2.3 Continued. Ascending from Earth.

map of Manhattan Island, showing the location of the museum. 10^6 – sheets of the new "international map," a famous project at the time that included the first complete map of the Arctic and Antarctica. 10^8 – a model of the Earth and the moon, showing the diameter of the Earth relative to distance to the moon, and models of the planets on the same scale. 10^{10} – a model of the Earth, moon and sun, with models of the satellite systems of the planets, and diameters of the largest orbits, up to a trillion, a model of the whole solar system. He only goes out to Neptune. Why does he stop at Neptune? Because Pluto hadn't been discovered. It was another 15 years before they discovered Pluto. Of course in the last few years, we've demoted it from its planet status. So he was actually right at the time, with the planets of the solar system ending at Neptune.

10^{-2} to 10^{-10}

Then he goes on down into the atom: 'Though it is outside my field, I can hardly refrain from adding the suggestion of a set of diagrams in the other direction.' A magnification of 100 and of 10 000 times would be registered in the field of microscopy. A million times would be ultra-microscopic particles. 100 000 000 times would illustrate molecular diameters and crystal structure. 10 000 000 000 times could perhaps illustrate 'Rutherford's nucleus atom', as he describes it.

Powers of 10 at the Hayden Planetarium

I've spent several years working on an exhibition space along the lines that were imagined by Henry Russell. We spent $210 million rebuilding the Hayden Planetarium at the American Museum of Natural History. The planetarium is in the form of a giant sphere, which stands within the Rose Center for Earth and Space (Figure 2.1). In the upper half, there is a space theatre, with a Zeiss fibre-optic projector simulating the stars in such clarity of detail that if you go in there with binoculars, you will see the stars even better than with your naked eye. But in the base of the sphere, you can see a whole different universe. We show the beginning of it all, the Big Bang.

10^{18} seconds ago

This is the beginning of time, 14 or so billion years ago. You step out of the Big Bang, and you take a walkway through time. We have laid down 14 billion years of cosmic time on a spiral 100 metres long. And on the walkway we've placed images of astronomical objects whose light hails from that time in the life of the universe. So at the 3-billion-year mark, you see a picture of an object whose light was emitted 3 billion years from the beginning and has been travelling ever since. Because when you look up at the night sky, you see the cosmos not as it is, but as it once was. Even light takes time to get to you across the universe.

10^{15} seconds

So every image shows an object from the history of the universe. Since it's a linear scale, one step taken by an average-sized visitor spans 70 million years. What does that mean? The dinosaurs became extinct 65 million years ago. So if I go to the end of this ramp, and take a single step back, that's when the dinosaurs became extinct. That's yesterday on the cosmic timescale. When did the dinosaurs arrive? They came in some 300 million years ago, which is a mere three steps beyond that.

10^{12} seconds

At the very end of the ramp, we've got to show civilization. We have mounted a single strand of human hair there. The left side of that hair was a trillion seconds ago. Cave paintings by cavemen. The right side of that hair is the modern day. All that we call human culture occupies the thickness of a human hair at the end of our ramp.

Space

The space around the sphere is not just the housing for the space theatre; the sphere is an exhibit in itself. Spheres are a fairly common shape in the cosmos, because the laws of physics conspire to give you spheres. So we've put a walkway around our sphere, and we give a powers of ten walking tour, comparing the sizes of many different objects and spaces across the size scales of the cosmos. There are a series of models mounted alongside the walkway, so that at each stage we can stand by the model, and say that if this model was expanded to be the size of the whole sphere, then some far smaller object – which can be seen in the next display along the walkway – would only be so big by comparison. One step at a time, visitors can imagine themselves walking through all these different scales (Figure 2.2).

10^{24} metres

We start with the observable universe. If the planetarium sphere was the size of the whole observable universe – including light that comes to us from 14 billion years ago – then relative to that size we can show the extent of space that contains the thousands of galaxies in the Virgo supercluster. In our exhibit, we show those galaxies as flecks within a solid glass ball right next to the walkway. Our own Milky Way is a member of the Virgo supercluster of galaxies. We can't look at the entire Virgo supercluster with a telescope, because we are in the middle of it, but we can see other superclusters around us.

10^{22} metres

If we made the planetarium sphere the size of the Virgo supercluster, a small globe on the railing would be big enough to contain our local group of galaxies. The Milky Way, the Andromeda galaxy, the Magellanic clouds, and a few other galaxies – our cousins, and brothers and sisters.

10^{20} metres

Now let's make our local group of galaxies the size of the planetarium sphere. Once we've done that, our own galaxy is on the scale of an exhibit by the walkway. So we live in a big family – the volume occupied by our family is large. The diameter of a galaxy is a hundred thousand light years. There's a (faint) picture of our galaxy in Figure 2.2. This is actually not a photograph, it's a constructed model of the light distribution of our Milky Way. It's very flat, and you can't see very far across because there are so many stars, but to see the rest of the universe, you can look out above or below the flat disk.

10^{18} metres

So now the planetarium sphere is the bounding volume of our galaxy, and we have a glass sphere the size of a baseball mounted on the railing of the walkway. This is a cluster of stars. We put about a hundred thousand specks in that ball, each speck representing a star. It's a globular star cluster. There are a couple of hundred of these clusters in our galaxy, and they orbit in big looping trajectories. Figure 2.2 shows an actual picture of a globular cluster that has about a hundred thousand stars. This is a photo taken by the Hubble Space Telescope.

10^{16} metres

Let's keep going. Now the planetarium sphere is that globular cluster of a hundred thousand stars. And another sphere slightly bigger than a cricket ball represents the volume of comets that orbit the sun – a volume of comets predicted to be there by the Danish astronomer Jan Oort. They comprise the Oort cloud. These are comets that come raining down on the inner solar system, but have orbits of tens of thousands, hundreds of thousands of years. You only ever see these once in a lifetime. Not like Halley's comet that comes around every 76 years. This is a cloud of comets. Trillions upon trillions of comets. And our solar system would be deep in the centre of this sphere. So now we're localized to the volume occupied by the gravitational influence of a single star, within the volume of that star cluster.

10^{14} metres

Now take that Oort cloud of comets, make that the sphere, and we have a little hockey-puck-sized exhibit. This contains our entire solar system – the orbits of all the planets. On the hockey puck we have a circle that is the orbit of Neptune. Then Uranus, Saturn and Jupiter. Mercury, Venus, Mars and the Asteroid Belt are in a tiny circle around the Sun. And there is another reservoir of comets, distinct from the far wider Oort cloud. This was predicted by Gerard Kuiper in the mid twentieth century, and it is known as the Kuiper belt of comets. Do you know who orbits out here among the Kuiper belt of comets? Pluto. Do you know what Pluto is more than half made of? Ice. Like all the other stuff out there. So Pluto has finally found a home. It's not just an oddball in the solar system, smaller than all the other planets, with this weird orbit that crosses the orbit of other planets. Do you know if Pluto was where the Earth is right now, the heat from the Sun would make it grow a tail. Now what kind of behaviour is that for a planet? So we've taken Pluto out of the pantheon of the planets, and put it with the comets. It's the biggest known object of the Kuiper belt. It's the king of the Kuiper belt! I think it's happier there, actually, being the biggest fish in a pond of small fish.

10^{10} metres

Now we make the planetarium sphere the Kuiper belt. And we have a little sphere on the railing of the walkway. This is the relative size of a blue super-giant star. They are called this because . . . they are very big and they are blue. The particular one that we've placed here is a blue super-giant called Rigel, the left kneecap in the constellation Orion. Two principal stars are his left kneecap, and his right shoulder area, Betelgeuse, a red super-giant star. Betelgeuse roughly translates as 'armpit of the great one'.

10^9 metres

So now, the planetarium sphere becomes a super-giant, and, by the walkway, we have the sun. The sun is tiny compared to some other stars, but one day the sun will grow to be a giant, and in so doing, it will engulf the orbits of Mercury, Venus and Earth. We'll be a charred ember, turning deep within its surface.

10^7 metres

When we make the sphere the sun, we have Mercury, Venus, Earth and Mars sitting on the rail, each smaller than the size of a soccer ball. They're not very impressive. They look like debris in the solar system compared to the size of the Sun. If the Sun were hollow, you could fill it with more than a million Earths, and have room left over.

10^5 metres

So now the planetarium sphere is Earth, and we show one of the moons of Saturn along the rail. Once the sphere is that satellite of Saturn, then we have a clay model of that famous crater in Arizona, made by a meteor – we call it Meteor Crater. If you played a football game in the bottom of Meteor Crater, you'd have seating for 2 million people around the rim.

10^3 metres

Now if the sphere is Meteor Crater, then on the railing we have a model of the asteroid that made that crater. The asteroid is very small compared to the crater. It came in with a lot of kinetic energy. That energy's got to go somewhere – some of it vaporizes the asteroid, the rest of it thrusts the Earth's crust out of that hole, and tosses it far and wide.

How about the crater responsible for the dinosaurs' extinction, 2×10^{15} seconds ago? It's centred on Yucatan, off the tip of Mexico, in the southern part of North America. The Yucatan crater is 200 miles in diameter, but the splash zone went as far as Minnesota, nearly a thousand miles north of there. So if you're a dinosaur, your choice would be either to stand right where the asteroid is going to hit, getting vaporised instantaneously, or go somewhere else: where it's raining fire; dust is thrown into the stratosphere; cloaks the Earth's surface from sunlight; knocks out the base of the food chain, and then you starve. I'd rather go quickly. It took out all the dinosaurs, no matter where they were on Earth.

10^0 metres

Here we have a one to one scale, which is kind of weird – you have to go through that at some point. So now, the sphere is the sphere. And our exhibit is a brain, the actual size of your brain. I like to pause here, and realise, as frail as we are in time and space, and as tiny as we are, and as small as our brain is, this brain actually figured all this stuff out. We're doing all right for ourselves.

10^{-4} metres

Let's keep going. Now the sphere is your brain, and we have a model which is an enlarged raindrop. Real raindrops are actually very small. Now let's go inside the raindrop, and see what can fit in there.

10^{-6} metres

Make the sphere a raindrop, and what we have is a red blood cell the size of your hand. Now you get a sense of how tiny cells are, compared with just a drop of water. Surely Anton von Leeuvenhoek was astonished putting the drop of pond water under his first microscope, and seeing a whole universe of creatures. You get a sense of how big, how much of a universe a single drop of water represents, if you're this small.

10^{-8} metres

Now make a red blood cell the size of the sphere, and we get the size of a virus along the rail, where we place a model of a rhinovirus, the cause of the common cold. Now you see how small viruses are compared with cells (Figure 2.2 shows the influenza virus). Most people do not appreciate how much tinier a virus is than bacteria and other cells. So the ways that you have to combat it in your body are very different from the ways you fight bacteria.

10^{-10} metres

Here's my favourite part of the entire journey. The rhinovirus is now the size of the sphere. And if the sphere is a rhinovirus, our display model shows the size of a hydrogen atom. What's remarkable to me is that when you construct a building out of bricks, those bricks are smaller compared to the building, than our model hydrogen atom is compared to the planetarium sphere. So imagine if we did have the tools to assemble atoms. It's not unrealistic to believe that we can create something like a virus from individual atoms, when we do something just as complex every day on a construction site.

10^{-12} metres

We're running out of universe now. We descend into the zone smaller than the electron orbitals that define the size of a hydrogen atom. We're inside the space of the atom and there are no known objects of this size. Atoms are so empty. There's a story about Ernest Rutherford. He was trying to find out how much space atoms take up, so he took thin gold leaf, and fired particles through it, to see how often they would slam into the stuff that was there. You'd assume that with solid material, particles would be hitting and bouncing off all the time. But nearly all of his particles went straight through, as though there was no gold leaf in front of them at all. Rutherford concluded, correctly, that atoms were mostly empty space. This was one of the first revelations in the world of modern physics. Rutherford, as a classical physicist, had to contend with that fact. It is rumoured that the next morning, when he woke up, he was afraid to step onto the ground, because he alone knew how empty solid matter was.

10^{-14} metres

Now we can show a model that represents the size of the uranium nucleus. It's one of the biggest nuclei of the periodic table, densely packed with 92 protons and nearly 150 neutrons.

10^{-15} metres

At this scale, if this sphere is now the size of the hydrogen atom that we saw earlier, then how big is a proton? We know it's going to be small. In our exhibit we drew a picture of it, accompanied by a caption on the picture. You know, even compared to the whole sphere of our planetarium, the proton is smaller than the dots that are on the "i" of its display caption.

10^{-18} metres

We know that electrons are smaller than this. So too are quarks. The proton is not fundamental matter. Quarks and electrons are, as far as we know. We still don't have a size measure for them. All we can say is that they are smaller than this measure.

Looking upward and outward

In our show at the planetarium, as we ascend from Earth, we see three other planets in the inner solar system, and then the sun. If we look very deeply into the night sky, in the Milky Way, we see an uncountable number of stars. Yet there are still no other galaxies in this picture, only our own galaxy.

In Figure 2.3 we see, across the plane of our galaxy, the hundred billion stars that are the Milky Way. Now, in our show, we ascend up out of the Milky Way, looking back toward it. This is the entire extent of stars whose existence we have visual confirmation of. Every star you can see in the night sky is contained within this volume, around us at the centre – in this random suburban corner of a spiral arm of the Milky Way. All the splotches are other galaxies, each containing a hundred billion stars of its own. Take another step back, and our own galaxy continues to recede into the distance. We can see our neighbour, the Andromeda galaxy. Our local group is here. Then we start to see the Virgo cluster coming into view. And we continue to ascend to the outer universe.

We enter the realm of one of the most famous images ever taken by the Hubble Space Telescope. It's called 'The Hubble Deep Field'. This contains some of the most fascinating structural information of galaxies ever recorded. Almost every smudge is a galaxy, as far as your telescope can see. Big ones, small ones, blue ones, red ones. Perhaps there are civilizations there, looking back to us, with a photograph that is a counterpart of ours. If we ascend a little farther out into space, whole clusters of galaxies are now nodes in a webwork of matter as it is distributed throughout the cosmos. The last part of figure 2.3 is a computer simulation of the clustering of galaxies as it appears in the universe. This is about a 10 per cent chunk of the total visible universe.

I think to myself: there was a time, in this expanding universe, where all the matter, all the energy, was contained in a volume the size of a proton.

That moment was the Big Bang. And so when I look at the cosmos from its beginning to its end, the very largest of scales owe their origin to something that was the size of a proton 14 billion years ago. So we have particle physics meeting astrophysics right back at the beginning of the cosmos. It's that perspective that I carry with me every day. When I look back from this image of the Universe and I ask 'where's Earth?' – back here in the Virgo cluster. Where's the Virgo cluster? Right here – these thousand galaxies. Where's the Milky Way? Somewhere in there. Where's Earth? Somewhere in that. On the scale of the Universe, we're lost among these powers of ten.

Let me leave you with one final thought: that the very chemistry of our bodies – the hydrogen, the oxygen, the carbon, the nitrogen – these elements are common throughout space. They are forged in the centres of high-mass stars that exploded and spread that enrichment throughout the cosmos. And from that enrichment, solar systems are born, and planets forged out of the debris – people, and life. And so yes, it's possible to feel small in the Universe, but I also feel large. Because I know that it's not as if I'm here and the Universe is there. It's not as though that is someplace else. It's not simply that we live in the Universe. It is also true, when we look across these 40 powers of ten, that the Universe lives in us.

3 The power of mathematics

JOHN CONWAY

This is a lecture about the power of simple ideas in mathematics.

What I like doing is taking something that other people thought was complicated and difficult to understand, and finding a simple idea, so that any fool –
and, in this case, you – can understand the complicated thing.

These simple ideas can be astonishingly powerful, and they are also astonishingly difficult to find. Many times it has taken a century or more for someone
to have the simple idea; in fact it has often taken 2000 years, because often the
Greeks could have had that idea, and they didn't.

People often have the misconception that what someone like Einstein did
is complicated. No, the truly earth-shattering ideas are simple ones. But these
ideas often have a subtlety of some sort, which stops people from thinking of
them. The simple idea involves a question nobody had thought of asking.

Consider, for example, the question of whether the Earth is a sphere or a
plane. Did the ancients sit down and think 'now let's see – which is it, a sphere
or a plane?'? No, I think the true situation was that no-one could *conceive*
the idea that the earth was spherical – until someone, noticing that the stars
seemed to go down in the West and then twelve hours later come up in the
East, had the idea that everything might be *going round* – which is difficult to
reconcile with the accepted idea of a flat earth.

Another funny idea is the idea of 'up'. Is 'up' an absolute concept? It was,
in Aristotelian physics. Only in Newtonian physics was it realised that 'up' is
a local concept – that one person's 'up' can be another person's 'down' (if the
first is in Cambridge and the second is in Australia, say). Einstein's discovery
of relativity depended on a similar realisation about the nature of time: that
one person's time can be another person's sideways.

Power, edited by Alan Blackwell and David MacKay. Published by Cambridge University Press.
© Darwin College 2005.

Well, let's get back to basics. I'd like to take you through some simple ideas relating to squares, to triangles, and to knots.

Squares

Let's start with a new proof of an old theorem. The question is: 'is the diagonal of a square commensurable with the side?' Or to put it in modern terminology: 'Is the square root of 2 a ratio of two whole numbers?' This question led to a great discovery, credited to the Pythagoreans, the discovery of irrational numbers.

Let's put the question another way. Could there be two squares with sides equal to a whole number, n, whose total area is identical to that of a single square with sides equal to another whole number, m?

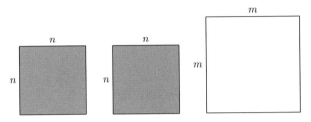

FIGURE 3.1 If m and n are whole numbers, can the two grey $n \times n$ squares have the same area as the white $m \times m$ square?

This damn nearly happens for 12 by 12 squares: 12 times 12 is 144; and 144 plus 144 equals 288, which does not *actually* equal 289, which is 17 times 17. So $17/12 = 1.416\,66\ldots$ is very close to $\sqrt{2} = 1.414\,21\ldots$ – it's only out by two parts in a thousand.

But we're not asking whether you can find whole numbers m and n that *roughly* satisfy $m^2 = 2n^2$. We want to establish whether it can be done *exactly*.

Well, let's assume that it *can* be done. Then there must be a *smallest* whole number m for which it can be done. Let's draw a picture using that smallest possible m.

Let's stick the two small grey squares in the top right and bottom left corners of the big square.

Now, part of the big square is covered twice, and part of the big square isn't covered at all, by the smaller squares. The part that's covered twice is shown in dark grey, and the bits that are not covered are shown in white. Since the

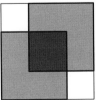

area of the original big white square is exactly equal to the total area of the light grey squares, the area of the bit that's covered twice must be exactly equal to the area of the bits that are not covered.

Now, what are the sizes of these three areas? The dark grey bit is a square, and the size of that square is a whole number, equal to $2n - m$; and the two white areas are also squares, with sides equal to $m - n$. So, starting from the alleged smallest possible whole number m, such that m^2 is twice the square of a whole number, we've found that there is an even *smaller* whole number $(2n - m)$ having this property. *So there can be no smallest solution.* Remember, if there are any solutions, one of them must be the smallest. So we conclude that there are *no* solutions.

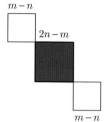

This result has tremendous intellectual consequences. Not all real numbers are the ratio of whole numbers.

This new proof was created by a friend of mine called Stanley Tennenbaum, who has since dropped out of mathematics.

Triangles

Take a triangle, any triangle you like, and trisect each of its angles. That means, cut each angle into three pieces, all the same size.

Extend the trisections until they meet at three points.

Then a rather remarkable theorem by Frank Morley says that the triangle formed by these points is *equilateral*. And this is true for any starting triangle.

Morley's theorem is renowned as being a theorem that's really hard to prove. Very simple to state, but very hard to prove. Morley stated the result in about 1900, and the first published proof didn't come till about 15 years later. However, I found a simple proof, aided by my friend Peter Doyle. Let me show you.

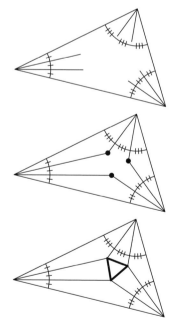

38

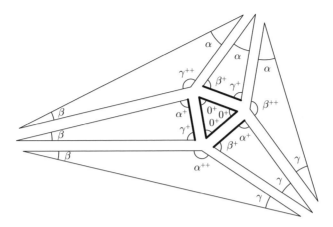

First, please tell me the three angles A, B, C, of your original triangle. Remember they have to add up to 180 degrees. Here's the plan. I'm going to *start* from an equilateral triangle of some size and build up six other triangles around it, and glue them together to create a triangle that has angles A, B, and C, just like yours; so for some choice of the size of the equilateral triangle, my construction will exactly reproduce your original triangle; furthermore the method of construction will prove that if you trisect your triangle's angles, you'll find my equilateral triangle in the middle. The diagram above shows the six triangular pieces that we will build around the equilateral triangle. This picture looks like a shattered version of the triangle we drew a moment ago, and indeed we'll in due course glue the pieces together to create that triangle; but to understand the proof correctly, you must think of the six new triangles as pieces that we are going to *define*, starting from my equilateral triangle, with the help of the values of A, B and C that you supplied. The previous page's figure is our destination, not our starting point.

We construct the six new triangles by first defining their shapes, then defining their sizes. To define the shapes of the six triangles, we fix their angles as shown in the diagram above. We define $\alpha = A/3$, $\beta = B/3$, and $\gamma = C/3$. We introduce a piece of notation for angles: for any angle θ, we define θ^+ to denote $\theta + 60$ and θ^{++} to denote $\theta + 120$. So, for example, the three interior angles in the equilateral triangle (which are all 60 degrees) may be marked 0^+. (You may check that the angles in each triangle sum to 180.) Next, we fix the size of each triangle that abuts onto the equilateral triangle by making the length of one side equal that of the equilateral. These equal sides are shown by bold lines on the first diagram.

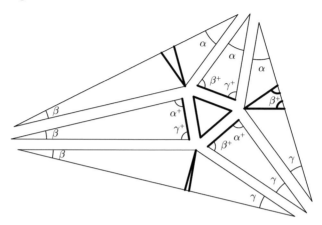

Next we fix the sizes of the three obtuse triangles; I'll show you how we fix the right-hand obtuse triangle, and you can use an analagous method to fix the other two. We introduce two lines that meet the long side at an angle of β^+ (a bit like dropping perpendiculars), and fix the size of the triangle so that both those lines have the same length as the side of the equilateral triangle.

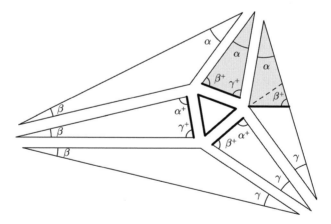

Now, having defined the sizes of all the triangles in this way, I claim that the two shaded triangles are identical – one is the mirror image of the other. We can see that this is so, because they have two identical angles (the αs and the β^+s); and they have one identical side (the highlighted sides, which are equal to the equilateral's side). Therefore the adjacent edges of those two triangles are identical in length.

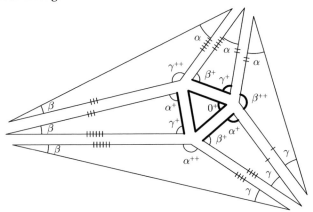

Applying the same argument six times over, we have shown that all the adjacent edges in the figure are identical to each other, and thus established that these six triangles will fit snugly around my equilateral triangle, as long as the angles around any one internal vertex sum to 360 degrees. The sum around a typical internal vertex is $\alpha^+ + \beta^{++} + \gamma^+ + 0^+$; that's five +s, which are worth 300 degrees, plus $\alpha + \beta + \gamma$, which gives a total of 360.

Thus, glueing the seven pieces together, I've made a triangle with your angles, for which Morley's theorem is true. Therefore, Morley's theorem is true for your triangle, and for any triangle you could have chosen.

Knots

Finally, I would like to tell you a little bit about knot theory, and about a simple idea I had when I was a high school kid in Liverpool many years ago.

First, what's the big deal about knots? Knots don't seem especially mathematical. Well, the first thing that's hard about knots is the question: 'Are there any?'

To put it another way, can this knot be undone? (It's conventional, by the way, to attach the two free ends of a knot to each other, so that the rope forms a closed loop.) The fact that no-one's undone it doesn't mean you necessarily can't do it. It might just mean that people are stupid. Remember, there are simple ideas that no-one had for 2000 years, then Einstein came along and had them!

Now, when we fiddle around with a piece of string, changing one configuration into another, there are three basic things that can happen. These are called the Reidemeister moves, after the German professor of geometry, Kurt Reidemeister. We'll call these moves R_1, R_2 and R_3.

R_1 involves twisting or untwisting a single loop, leaving everything else unchanged.

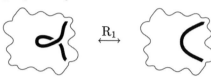

R_2 takes a loop and pokes it under an adjacent piece of rope.

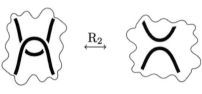

R_3 is the slide move, which passes one piece of rope across the place where two other segments cross each other.

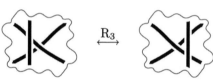

All knot deformations can be reduced to a sequence of these three moves.

Now, is there a sequence of these moves that will enable you to start with the knot on the left and end with the 'unknot' on the right? You

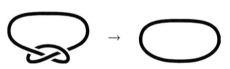

can perhaps imagine applying a sequence of moves until it really looks rather messy – imagine a picture like this, but with maybe a million crossings in it:

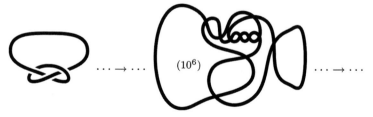

And maybe eventually, if I'm lucky, another million moves would bring me to the unknot.

Can you disprove this story?

It is quite hard to disprove it. I believe no one has ever tried going out to all the mindbogglingly large number of million-crossing configurations and checking what happens in each case. And the difficult challenge is, if we want to prove that *a* knot exists, we must show that *no* such sequence of moves exists.

What I'm going to do is introduce what I call a *knumbering* of knots. To make a knumbering, you assign a little number to any visible piece of string; and in a place where one piece disappears under another, the two numbers associated with the *lower*

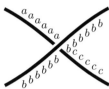

piece of string must be related to each other in a way that depends on the number on the *upper* piece of string. Namely, if the number on the upper piece is *b*, and the lower piece's numbers on either side of the upper piece are *a* and *c*, then '*a, b, c*' must be an *arithmetic progression*. That means the amount by which you go up from *a* to *b* has to be exactly the amount by which you go up from *b* to *c*.

For example, if *a* is 13 and *b* is 16, then *c* had better be 19.

Now, what is the relevance of these numbers? Well, let's first see if we can make a knumbering. Let's take our old friend, the trefoil knot, and work our way round the

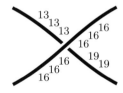

knot, assigning numbers to its different segments, and see if we can satisfy the arithmetic progression condition at every crossing. How should we start? One thing worth noticing about the arithmetic progression condition is that it is *invariant*: I can shove all the numbers *a, b* and *c* up or down by any amount I like, and they will still satisfy the arithmetic progression condition. So we may as well start by assigning the labels 0 and 1 to a couple of edges here, then we can propagate the consequences of those choices around the rest of the knot. We'll mark each crossing as we apply its rule.

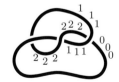

So far, so good. . .

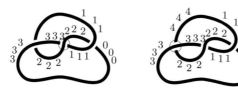

Oh dear, there is a problem on the top edge, namely that 4 isn't equal to 1, and they should be equal, because there's a 4 and a 1 both on the same piece of rope. However, one of the great powers of the mathematical method is that I can define things however I like; so I'm now going to define **4** to be **equal to 1**. (Mathematicians call this kind of equality 'congruence modulo 3'.) So, phew! I cured that problem.

There is a similar contradiction on the bottom segment: this edge is labelled both '3' and '0'. But if 4 is equal to 1, then 3 is equal to 0. So everything is all right.

We have got a knumbering.

Now, what's the point of these knumberings? It is very beautiful. Look at what happens when we take a knumbered knot and apply the three Reidemeister moves to it. Can we take the left-hand knumbering and obtain a right-hand knumbering?

The answer is yes, any valid knumbering for the left-hand figure can be copied into a valid knumbering for the right-hand figure, and vice versa. This is quite easy to confirm for the first two moves.

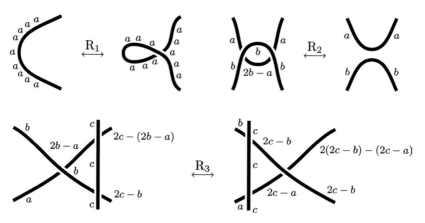

For the third move, we have to confirm that $2(2c - b) - (2c - a) = 4c - 2b - (2c - a) = 2c - (2b - a)$.

We find that we can do any of the three moves without messing up the rest of the knumbering.

The fact that any valid knumbering remains a valid knumbering when a move is made or unmade implies that the *number of possible knumberings* of the left-hand picture is exactly equal to the *number of knumberings* of the right-hand picture.

Now, let's return to the question of whether the trefoil knot can be transformed into the unknot. Well, there are just three knumberings of the unknot.

Whereas the trefoil knot has at least four knumberings: the all-0, all-1, and all-2 knumberings, *and* this one.

So now, we can prove that *the trefoil knot cannot be undone*, because it has a different number of knumberings from the unknot. If the trefoil knot and the unknot were related by a sequence of Reidemeister moves, they would have the same number of knumberings.

This proves that knots *do* exist.

Tangles

I often do a little conjuring trick which consists of tying knots. Tangles are bits of knottiness with four ends coming out, and they have an unexpected connection to arithmetic.

Tangles are best displayed by four square-dancers. Two dancers hold the ends of one rope, and two dancers hold the ends of the other rope. We can manipulate the tangle by using two moves, called *twist 'em up* and *turn 'em roun'*.

When we *twist 'em up*, the two dancers on the right-hand side exchange places, the lower dancer going *under* the rope of the upper dancer. Now, we're going to assert that each tangle has a value, and that 'twist 'em up' changes the value of the tangle from t to $t + 1$. (These values aren't related to knumberings; you can forget about knumberings now.)

twist 'em up

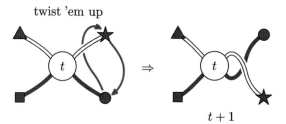

$$t + 1$$

When we *turn 'em roun'*, all four dancers move one place clockwise. 'Turn 'em roun' changes the value of a tangle from t to $-1/t$.

turn 'em roun'

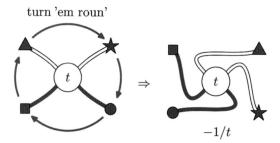

$$-1/t$$

To get us started, the tangle shown below is given the value $t = 0$.

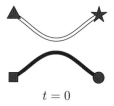

$$t = 0$$

Is everything clear? Then let's go!

twist 'em up

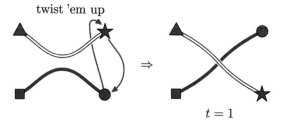

$$t = 1$$

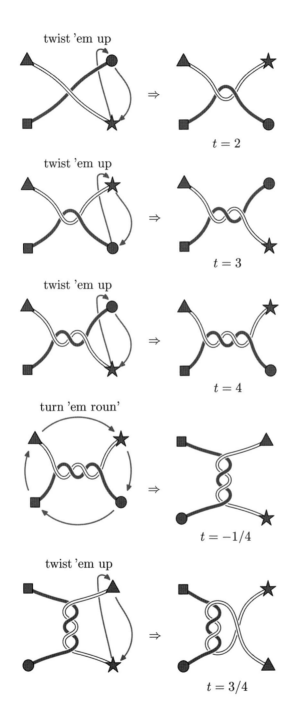

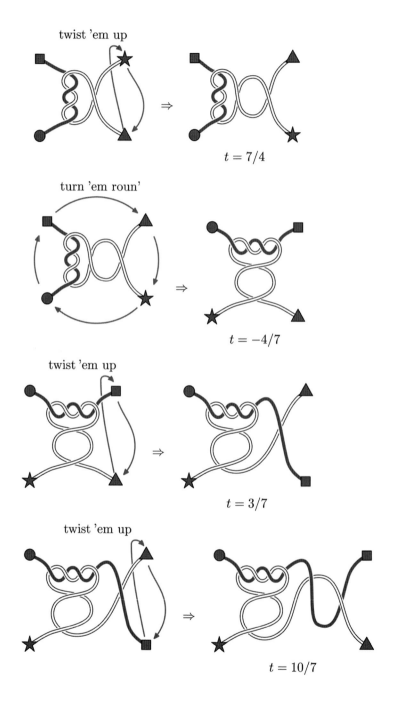

twist 'em up

$\Rightarrow$

$t = 7/4$

turn 'em roun'

$\Rightarrow$

$t = -4/7$

twist 'em up

$\Rightarrow$

$t = 3/7$

twist 'em up

$\Rightarrow$

$t = 10/7$

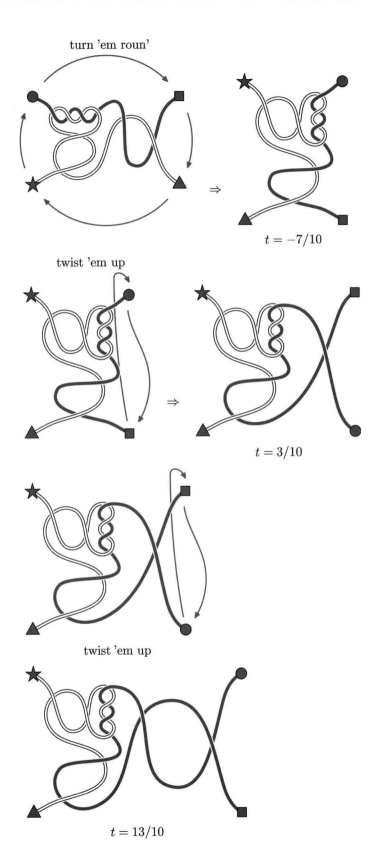

turn 'em roun'

$t = -7/10$

twist 'em up

$t = 3/10$

twist 'em up

$t = 13/10$

Now, it is your job, dear reader, to get the dancers back to zero. But you are only allowed to do the two moves I've spoken of. Do you want to twist or do you want to turn?

What you'll find is that if you use your knowledge of arithmetic to get the value back to zero, *the tangle will indeed become undone*. It's magic! [The sequence chosen by the audience in Cambridge was: $13/10 \xrightarrow{r}$ $-10/13 \xrightarrow{r} 3/13 \xrightarrow{r} -13/3 \xrightarrow{u} -10/3 \xrightarrow{u} -7/3 \xrightarrow{u} -4/3 \xrightarrow{u}$ $-1/3 \xrightarrow{r} 3 \xrightarrow{u} 4 \xrightarrow{r} -1/4 \xrightarrow{u} 3/4 \xrightarrow{r} -4/3 \xrightarrow{u} -1/3 \xrightarrow{u} 2/3 \xrightarrow{r}$ $-3/2 \xrightarrow{u} -1/2 \xrightarrow{u} 1/2 \xrightarrow{r} -2 \xrightarrow{u} -1 \xrightarrow{u} t = 0$, with twist 'em up and turn 'em roun' abbreviated to $\xrightarrow{u}$ and $\xrightarrow{r}$, respectively.]

This is an example of a very simple idea. We already knew some arithmetic – but only in the context of numbers; and we didn't realise it applies to knots. So in fact this little branch of knot theory is really just arithmetic.

Having found this unexpected connection, let's finish with something fun. Start from $t = 0$, and turn 'em roun'. What do we get?

turn 'em roun'

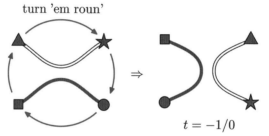

$t = -1/0$

Hmm! Now we've got $-1/0$, isn't that some sort of infinity, or minus infinity?

Let's see what you get when you add one to infinity. Does adding one to infinity make any difference? Twist 'em up!

twist 'em up

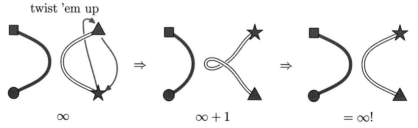

∞ $\qquad\qquad$ $\infty + 1$ $\qquad\qquad$ $= \infty!$

Isn't that nice? We add one to infinity, and we get infinity again.

So, this is a powerful idea that we mathematicians use: you take something you've learnt in one place, and apply it to something else, somewhere where it's not obvious that there is any mathematics, and there *is*.

4 The power of narrative – 2-D/3-D/4-D

MAUREEN THOMAS

Is all the world a movie, and are all the men and women in it merely players? In the twenty-first century, in the Western world and many other parts of the globe, the screen is as widespread a medium for narrative as the page was in the nineteenth, or the stage in the sixteenth century. The miniaturisation and accessibility of audiovisual equipment enables people to use moving images much as earlier generations used paper, pencil and pen to record and tell stories. In fact, there are a hundred years of history behind the development of contemporary screen language, and screen literacy is a refined art. How does it work to engage and affect us? Exemplified through close reference to movies and interactive titles from 1922 to 2002, this examination of the workings of screen narrative follows the evolution of the fictive world beyond the 2-D frame from cinema to computer screen – from 2-D to 3-D to 4-D.

Though often featuring actors playing roles, film drama is recorded and delivered through lenses, the camera representing a single viewpoint, mediating the story via an inbuilt observer, narrator or character. Film-makers thus wield the narrative powers of both novelists (who control viewpoint and narrative stance absolutely in their writing) and dramatists (who offer the interplay between characters, observable from a number of perspectives, as the vehicle for narrative) to spellbind their audiences. But in addition, like the nineteenth-century illusionist and pioneer film-maker Georges Méliès (1861–1938), who first recognised the potential of film as magical entertainment, moviemakers can manipulate images to make us actually see, and believe (rather than merely imagine), the fantastic and the improbable. The advent of interactive digital technology, virtual cameras (which can adopt a theoretically unlimited range of viewpoints) and animated performers (each of which may incorporate the skills

Power, edited by Alan Blackwell and David MacKay. Published by Cambridge University Press.

51

of many live performers, ranging from acrobats through martial arts experts, dancers and singers to traditional dramatic actors) increases the strength of an already potent mix. In the twenty-first century, the power of screen narrative attains new heights.

Traditional 'Epic' offers a hero with whom storyseekers can identify through an imaginative fictive effort, usually learnt as a child through personal experience of storytelling and make-believe play forms. In the twentieth century, moving image narrative (television, film, computer and console games) was the medium through which many, if not most, young people learnt to identify dramatic characters, and identify with their fictive experiences. Typically, in these media, an epic hero traverses an expressive narrative landscape which can both portray a physical place (external quest) and represent a psychological landscape (internal quest) – exemplified in cinema by Dorothy in the *Wizard of Oz* (Victor Fleming, USA 1939), Paul Hackett in *After Hours* (Martin Scorsese, USA 1985), Neo in *Matrix* (Wachowski Brothers, USA/Australia 1999), Lara Croft in *Tomb Raider* (Simon West, UK 2001) or Aki Ross in *Final Fantasy* (Hironobu Sakaguchi, Japan 2001).

Epic patterns usually offer a singular imaginative experience, where events, settings and characters are wholly or mainly seen from the viewpoint of the hero, and are important only as they affect him or her. The experience of the hero becomes the imaginative experience of the viewer, reader or listener, who, for the duration of the fiction, enters a state where external stimuli are excluded or ignored, and the life of the imaginary world becomes temporarily more immediate and real than the life of the everyday physical world. In the rhetoric of moving image language, music, sound, colour, rhythm, spectacle and the actual kinesis of the image-flow itself play a significant part in submerging the storyseeker in the world of the fiction. At the same time, they supply the stimulating and satisfying aesthetic experience which make cinematic, televisual, animation and video adventure-games such engaging forms of pleasure.

In the twenty-first century, the realm of film and analogue media has extended to include computer-based digital media. The dramatic and narrative power released by this development allows a fusion of the creative talents of storyteller, dramatist, architect, visual artist, software-designer and computer-programmer, to generate engaging and satisfying storyforms. The traditional skills of set designers, cinematographers and animators in dramatising landscape and setting, need, in the 3-D animated navigable world of the *Tomb Raider* (Core Design / Eidos Interactive 1996–2002) or *Final Fantasy* (Squaresoft / Squaresoft 1995–2002) interactive console adventures, to amalgamate with the traditional talents and experience of moving image fiction directors, for effective, involving narrative drama to be produced in these modes. A creative synthesis between what in film and television is treated as the *location* or *set*, the *action* of moving-image fiction and *virtual cinematography* creates new narrative conventions and possibilities for narrative stance. As, in navigable 3-D storyspace, the role of the passive consumer transforms into that of the participative storyseeker, the environment of dramatic adventure beyond the frame is enlivened into an active architecture of story*telling* and of story*seeking*.

In Larry Semon's silent movie *The Wizard of Oz* (USA 1922), based on the ('children's') book by L. Frank Baum, young Dorothy's wooden house is blown across the screen from Kansas to the 'Land of Oz', pursued by a silent comic farmhand (who later becomes the Cowardly Lion which, in the adventure, goes in quest of Courage to the Emerald City), until it lands in the dream country of Oz. As the house sails crookedly across the sky, the farmhand, trying to catch up, magically runs through the air, driven by vivid streaks of lightning, which make him throw up his hands and run faster every time a flash connects with the seat of his pants. The screen offers a flat space, across which a flat prop-house sails in horizontal motion, as the farmhand runs in silhouette like a shadow-puppet across the round white platter of the full moon. There is no sense of depth at all – the illusion of flight is created against a backdrop which looks like a painting come to black-and-white life.

The flat white rectangle of the cinema screen is an integral part of the show. It is the plane upon which the flickering lightbeam of moving-image magic can project its insubstantial and ephemeral phantoms; a fragile membrane, where, for a moment, the visions of imagination and dream can be conjured into fantastic life. The audience of this brand of 1920s cinema – which is close kin to the tricks and illusions of its cradle, the fairground show, and is not far from the displays of mediums or spirit-raisers who performed their arcane rituals in the half-dark – could enjoy the thrill of seeing unfold before their eyes something that common sense told them could not possibly be happening: they could admire the art with which the trick was achieved, even as they were thrilled by the invention of the artists' minds, and delighted by the visual splendours of the movie-set world the silver screen displayed.

In Victor Fleming's much better-known Technicolor sound movie, *The Wizard of Oz* (1939), on the other hand, when Dorothy's home is blown from Kansas to the 'Land of Oz', the audience (that is, the camera) is with Dorothy inside the 3-D space of the 3-D house, as it spins through the air, deeper and deeper into the screen. As Dorothy knocks her head, and falls onto the bed, the subjective camera-work pulls the picture in and out of focus, so that the audience, rather than watching an illusion from which it can stand back in admiration, is obliged to share Dorothy's experience of dizziness and faintness, drawn into the fictional world by the director's control of cinematic viewpoint and narrative stance. However, when (dreaming) Dorothy raises her head to look out of the cabin window at the objects flying past in the hurricane, the window-frame acts, for her and for the audience, very like the old 2-D silent cinema-screen rife with inventive illusions – the airborne schoolmistress/witch skims by on her bicycle with Dorothy's confiscated dog, Toto, in her basket; the farmhands (later to become Dorothy's questing companions, the Straw Man, the Tin Man and the Cowardly Lion) row desperately through the clouds in a small dinghy – all whiz across the flat space of the window/screen just as Semon's comic farmhand did in 1922. In this framework, 'Screen' and 'Dream' are one and the same. Fleming offers a '2-D cinema' within a '3-D cinema', and, like the Shakespearian device of the play-within-a-play (e.g. *Hamlet*, *Midsummer's Night's Dream*, *The Tempest*), this has the effect of giving the audience the sense that the action of the frameplay, however improbable, is somehow real, because the play within it is, by contrast, so obviously contrived or magical.

In Fleming's 1939 movie, the illusory world inhabited by Dorothy is a solid 3-D set, and, to the audience, the screen on which the movie is projected is made to seem more substantial both by the 3-dimensionality of the set, and in comparison to the even more fragile dreamframe of the cabin window within the film. In this version of the movie, as Dorothy's house falls dizzily out of the sky directly towards the audience, the 'special effect' of the tiny 3-D model house spinning through the endless ether compels spectators into the illusory world on the other side of the frame, because the director, controlling the viewpoint and dramatising the action, forces them to take the position of the Wicked Witch herself, upon whom (we soon discover) the house so crucially lands, crushing out her life. Killing the witch makes Dorothy an instant hero, and sends her on her quest, so this is a climactic moment for the story. When the house lands with a thump emphasised both by the camera strategy and by the sound-effects, which are followed by an expectant hush, the camera, inside the cabin, moves to follow Dorothy outside – into the colourful new world of Oz (designed, with all the brilliance of a Broadway musical stage, but with deeper colour and more dazzling detail, by Cedric Gibbons). The camera takes Dorothy's point of view of the many-hued land of the Munchkins, as the black-and-white movie-world in which the Kansas section of the film takes place is left behind with the wrecked old house, a moment orchestrated by the soaring melody of the song, 'Somewhere over the Rainbow'. The hyper-'real' world of Oz, the magic land on the other side of the screen, is not only more attractive, colourful – wonderful – than Dorothy's Kansas: it is brighter and more beautiful than the everyday world of the audience. The audience shares the camera's, and therefore Dorothy's, point of view as the wondering child beholds this universe of enhanced colour and sound for the first time. The adventure begins with a swooping movement of the camera into the luscious landscape, as Dorothy walks tentatively through the 3-D set, fertile with oversized, over-coloured flowers. It is a garden of romanticised childhood memory, viewed as by small people, where nature is vast, rainbow-coloured and replete with wonder. The camera moves with Dorothy – like the child stepping through the doorway from the world of black-and-white 2-D movies into the 3-D Technicolor Land of Oz, the audience has the sense of passing through the frame of the cinema screen, and entering into the 'real' 3-D world behind it.

The ability of camera movement, colour and sound to transport an audience seated in a large dark auditorium into a world beyond the screen is exploited more and more expertly as the twentieth century progresses, and more mobile cameras, more sensitive filmstock and more sophisticated sound recording, mixing and playback devices develop. In *After Hours* (Martin Scorsese, USA 1985), the hero, Paul Hackett (who, as his name suggests, spends his waking hours in vast, impersonal, noisy office spaces, training ambitious young men to use computers to leap over his head and achieve their career goals) has an adventurous night in SoHo, New York, after leaving the office. Scorsese's Paul does not, like Fleming's Dorothy, receive a physical blow to the head which catapults him into the dreamworld of Oz – he merely relaxes on the austere couch in his bland impersonal apartment before the flickering screen of the TV, remote control in hand. A cinematic cut to the all-night diner and coffee-bar where the definitive encounter of the night takes place leaves doubt as to whether Paul is sleeping or waking. Whichever it is, Paul takes a New York yellow taxi to visit alluring young Marcy at her apartment, ostensibly to buy a plaster bagel from her sculptor room-mate, Kiki. The camera observes Paul as he hails the cab and gets in – lingering, as it zooms from the kerb, on the black-and-white chequered stripe along the taxi's door, which seems to flow like the yellow brick road Dorothy has to follow to find the object of her quest.

Paul is whisked along by a taxi-driver whose maniacal style of negotiating traffic is exaggerated by speeded-up film and frenetic Spanish music (apparently from the cab radio) as the audience shares Paul's nightmare ride into the strange territory of after-hours SoHo. A subjective shooting-strategy – not looking at the world from Paul's point of view, but seeing Paul in his world (the interior of the taxi) through a violently swinging camera, which responds to every swerve of the cab as Paul is thrown from corner to corner of the back seat, so that the audience experiences a visual analogy to Paul's terrifying ride, while the music assaults their ears and nerves just as it may be supposed to assault Paul's – forces spectators to share Paul's experience as the taxi rockets at improbable speed through New York. The camera itself is fluidly dynamic, placed well inside the 3-D location-world of the screen. Unlike the larger, less manoeuvrable, studio-bound camera of the *Wizard of Oz* in 1939, it can be inside the cramped cab with Paul; and the New York through which the cab flies is not a studio set, but a real place. The camera penetrates and navigates the location, mediating Paul's experience through a visual language possible only on the screen. Paul's despair as his last dollar bill nightmarishly floats from the taxi before he can pay his fare is reflected in the horror on his face – a rapid close-up allows the audience to see and understand Paul's feelings – before a cinematic cut obliges them to share his actual point of view, as the irrecoverably lost note spirals gently down through the night sky, on its improbably long trajectory towards the earth. The dance-like rhythm of the music, a desolate flamenco guitar phrase, emphasises this protracted moment of the note's fall.

The whole shot is in fact set up just like the shot of Dorothy's house spiralling to earth in Fleming's *Wizard of Oz* – and if the combination of Paul's dizzying yellow taxi-ride from his dreary day-world into the adventureland of night-time SoHo with the spiralling bill falling from the sky fails to evoke a sense of Oz, in the café where he goes with Marcy (whose allure summoned him on this 'after-hours' quest, which will finally lead him home to himself), she earnestly explains that her ex-husband was obsessed by the movie *The Wizard of Oz*, even crying 'Surrender Dorothy!' every time he reached sexual climax. Not surprisingly, as Marcy makes this revelation, Paul begins to feel increasingly nervous of her neurotic personality. This nervousness is conveyed to the audience not just by the quality of the dialogue, and Rosanna Arquette's un-nerving performance as Marcy, but also by the camera-strategy, which uses close-ups to show Paul's reactions to Marcy's words, and reverse-angles from Paul's point of view of Marcy in close-up to convey his startled impression of her unsettling demeanour. The audience is invited to share Paul's subjective experience of the dramatic situation, as he begins his long nightmare quest to get home. New York at night becomes an eerie dreamspace, as the wet, dark, streets reflect wavering coloured lights into a screenspace where no pedestrian or vehicle appears, unless directly involved in Paul's adventure. SoHo is a ghost city, populated only by the archetypal spectres of Paul's own fears.

Scorsese's story (written by Joseph Minion) takes the spine of Dorothy's dream magic adventure – her quest to find all the things she needs to become a confident, self-reliant person, and get back home to Kansas – and transposes it to the world of the late twentieth-century office-worker/computer nerd, who has to face his innermost fears (of women and castration), before he can be 'reborn' into his home world, as a free individual. The tale of a quest, carried on through a half-real, half-dream psychological and physical landscape, offers a very powerful narrative dynamic, and one with which it is easy for most people to identify. With the aid of the precise and manipulative use of camera and sound, the cinema screen can become a penetrable membrane, through which the audience – cued by colour, sound, movement and dramatic performance – is carried on the wings of imagination into the dreamspace/storyworld of the actors in the drama. The feeling of being in insecure, threatening surroundings, driven by the need to find the way home, is the stuff of many an archetypal dream, and offers a situation everyone has experienced in some degree in waking reality. Both *The Wizard of Oz* and *After Hours* provide a powerful narrative matrix, where the hero (with whom the audience identifies) passes through trials which develop the strengths, skills and understanding that finally enable her or him to complete the task(s) or achieve the goal(s) set in the epic story, negotiating the expressive space of a dreamworld echoing with reverberations of reality.

Steven Spielberg, in *ET* (USA 1982), chooses as protagonist a creature with whom both dramatis personae and audience might have trouble identifying – an Extra-Terrestrial alien. This is very much the point of the story – hero Henry Thomas, the boy who finds the stranded ET, has to develop empathy with a creature he does not understand, and finally recognise its right to go home to its planet, and not merely serve as an object of (possibly destructive) curiosity to those able to capture it. How does Spielberg create a corresponding empathy in the audience? From the opening sequence, where forest ferns appear gigantic, and sound and light are unbearably intense, the camera identifies with ET's point of view – the small alien itself does not appear on screen, but the audience experiences its sensations, because the camera moves, at ET eye-height, through a threatening, but also magical, night-time forest (very like the nightmare wood surrounding the Wicked Witch's castle in Fleming's *Wizard of Oz*), and the sounds of the humans who pursue the little creature, invasive torchbeams dazzling, keys jingling on keyrings with preternatural loudness, assault the audience from all sides through the surround-sound system of the movie theatre.

Movie sound can very powerfully create subjective identification between audience and onscreen character. As Scorsese used the frenetic music inside the taxi to pull the audience into Paul Hackett's nightmare situation ('source music', apparently coming from somewhere in the picture, in this case the cab radio), so Spielberg uses exaggerated sound-effects coupled with John Williams' highly dramatic score ('film music', coming from outside the action of the movie, heightening the emotional impact of the story) to plunge the audience straight into ET's plight at the very start of the film. Spielberg's camera-strategy does not merely show the character in such a way as to evoke feelings in the audience parallel to those of the character (as Scorsese's does, where the violent swinging motion of the camera inside the taxi interprets Paul's disorientation as the audience watches him swept through his terrible cab-ride) – the camera itself becomes ET's experience, lurching through the undergrowth as ET hurries to try to board the spaceship which carries his companions away from the pursuit of the large and threatening Earth-dwellers. ET's first clear appearance on screen is as a pair of gnarled fingers, parting the fronds to watch the ship, silhouetted against the silver disk of the moon, as it leaves, abandoning him to his fate. The audience experiences this abandonment at first hand – only ET can see his own fingers from the angle shown by the camera, so the audience literally sees what ET sees, the camera transposing them into ET's position. The audience shares ET's viewpoint absolutely, and at the same time is immersed in the soundtrack, just as ET is overwhelmed by the frightening soundscape of the alien forest. Sound in the cinema can literally surround the audience, as the flat 35 mm screen can never do, and Spielberg controls narrative stance through the combined power of sound and camera-strategy, from the outset melting away the frame of the storyspace (the cinema screen) into darkness, to draw the audience directly into the fantasy world of the movie, on the other side.

The audience experiences ET's adrenaline rush as he flees the humans who pursue him, empathising not with the humans (which is what the spectators actually are), but with the alien; seeing humanity as ET sees it: threatening, arrogant, powerful, aggressive towards anything it doesn't recognise – and dangerous. The action moves fast, and because we have no alternative but to see the world as the camera, directed by the director, sees it, we shift our identity, for the duration, sliding imperceptibly into the persona of the little Extra-Terrestrial. Through controlled use of powerful subjective camera and sound strategies, the 'us' of the screen world have become 'them', and the 'them' have become 'us'. Perhaps the most resonant image of the movie is that of ET expressing his desire to contact his own people, holding up those two gnarly fingers, and speaking the human words, 'ET phone home'. Like Dorothy in *The Wizard of Oz* and Paul in *After Hours*, ET is the archetypal lost one, alone amongst creatures not of his own kind, trying desperately to find the way home.

Cinema, like the novel, can adopt different narrational stances, and can change them within the same movie. In ET, the camera starts by identifying us with protagonist ET, in a first-person narrative. Later, we identify also with hero Henry, who determines to help ET get home. In the final dash for ET's freedom, Henry and his group of friends, with ET on the carrier of Henry's bike, cycle frantically across suburbia to get the alien to the ship he has suc-ceeded in contacting and drawing down to pick him up. The boys on their bikes desperately race the police in their squadcars – finally evading them by lifting magically off into the sky. At this moment, the kids on their bikes appear to fly straight out into the audience, and then sweep on overhead, as the camera-work and editing, emphasised by the swelling chords of the score, create the illusion that the space in the screenworld actually extends out into the auditorium – the proscenium is inadequate to contain it. This mise-en-scène (action staged for the camera) doesn't just allow the audience to enter the 3-D world beyond the screen, it drags the viewer willy-nilly into the action by using the camera and editing to throw the environment of that action out to envelop the audience, projecting the illusion through and past the screen. Then the camera switches to the subjective points of view of the boys themselves as they soar above the earth, and the audience flies with them. For a moment, Spielberg cuts back to show them flying in a skein across the great round face of the setting sun, in a shot echoing those that took Dorothy to Oz in 1922, and again in 1939; but in 1982, with the help of sophisticated cinematic 3-D special effects, the audience are not just watching – they feel they are participating in the adventure, like the boys, flying through the sky.

In 1999, the Wachowski Brothers' *Matrix* (USA/Australia), made after the advent of the 3-D animated computer-game, and the development, during the 1990s, of sophisticated digital effects which permit computer-animated, manipulated and generated images to blend seamlessly with filmed action and location-shooting, shows computer-hacker Thomas, aka Neo, taking his first trip into the 'Real' world of virtual gameplay. In the story, mentor Morpheus inducts Neo into an understanding that the everyday world he inhabits, or seems to inhabit, is an illusion – the norm of corporate existence is a hallucination, devised, induced and sustained by the aliens who feed on human energy as a spider does on living insect prey.

Neo's transit from one world to another is the reverse of Dorothy and Paul's dream journeys: where they are spun by tornado or taxi from a waking world to a dream one, Neo moves from a tranceworld, a shared 'virtual' simulation which has the appearance of waking reality, through his own body/mind to the vision of 'reality' where huge, menacing, arachnid creatures manipulate comatose human forms cocooned in something resembling amniotic sacs penetrated by pink tubes, hinting of some vast, perverted weblike life-support system. For the trip, digital technology turns the screen into Neo's 3-D 'interior', and the audience spins through it, spiralling down a gullet/gutlike tunnel into the screen and Neo's mind. Spielberg in *ET* flung the screenworld out beyond the proscenium into the auditorium; the Wachowskis use virtual camerawork, where the 'camera angle' can be situated anywhere – since there is no physical camera, only digital images generated and manipulated directly to give whatever viewpoint the director chooses – to pull the audience deep into the tunnel of the screen. The space on the 'other side' of the membrane is not just a dream world, it is an interior world. The audience passes literally through the outward-seeming and physical body of the hero into the landscape inside his head.

Nonetheless, the basic scenario in *Matrix*, right at the end of the twentieth century, is the same one Semon used near its beginning, in the twenties, followed by Fleming in the thirties, and by Scorsese and Spielberg in the eighties. As Neo is strapped into the electronic chair which will catapult him into 'real' reality, 'Fasten your seat-belt, Dorothy', remarks Morpheus' assistant (soon to be revealed as the villain), 'Cos Kansas is going bye-bye'. As in *After Hours*, where the reference to *The Wizard of Oz* is overt as well as covert, the Wachowskis pay open homage to a movie that inspired them. In this adventure, hero hacker Neo is the 'One' who, aided by sidekick Trinity and guided by mentor Morpheus, is to save the World from the predatory aliens feeding upon humanity, by penetrating to the core of the fantasy, and fighting the enemy with its own delusive weapons. The cinema screen glows like the monitor of a computer, for which it sometimes explicitly stands, displaying numbers and letters in lurid green pixellated patterns. The audience looks into a virtual gameworld, where the rules of the digital domain apply – such as the animation trick of 'morphing' your shape, and the ability of 'players' to change character-avatars and appearance in mid-game, or jump instantly from one virtual timespace location to another.

This screen world is not the flickering projection of the 1920s onto the fragile illusionist's screen, a trick with light and shadows to make the audience gasp with delight at its consummate artifice; nor yet is it the more complex double illusion practised by Fleming in the thirties, as he plays on the cinema-screen within a cinema-screen to create an arena where an adult moral tale can be enacted in the dream–circuslike environment of childhood. It is not transmitted through Scorsese's sleight-of-camera and soundtrack approach, which for the duration of the performance banishes consciousness that this is a screen world at all, transforming the real SoHo location, by the cunning use of lighting and the weird depopulation of an ever-populous place, into a convincing waking nightmare. Nor is it Spielberg's immersive illusion of childhood, where sophisticated cinematic rhetoric plunges the audience inside the fight for freedom and justice against a desensitised adult conspiracy to drive out magic, represented by Henry's crusade to save ET – releasing adults and children alike from their own day-to-day selves into the persona of the little lost alien, seeking companionship, kinship and home. In *Matrix*, the big screen borrows the attributes of a computer-generated virtual environment, using the go-anywhere 'virtual camera' to create a narrative stance impossible in live-action movies before the advent of complex digital effects. Not only can the design, camerawork, editing and soundtrack create the illusion that the audience penetrates beyond the big screen into the cinematic world; the use of angles of view generated in the virtual environment of character-centred computer-games, which use 3-D animation in navigable storyscapes, allows the Wachowskis to go beyond cinematic rhetoric, to simulate an immaterial, totally navigable, penetrable cyberworld beyond the screen. The movie still exploits the traditional allure of the well-done conjuring trick fundamental to cinematic fantasy, but using digital effects: Trinity, a real flesh-and-blood actress, runs round three dimensions of a room, leaps the gaps between the roofs of skyscrapers, and moves through timeframes where a bullet can hang frozen in timespace – all before our very eyes. Dorothy closed her eyes, wished hard enough, and tapped the heels of the red slippers together to traverse spacetime; Paul Hackett travelled on the wings of repressed fantasy; ET had to use a spaceship – but Neo travels in digitality, a world where electrons and silicon chips provide the magic transport.

The idea of virtual characters playing out fantasy adventures is certainly not an invention of the Wachowskis – it has been familiar since the late 1980s to the millions of international fans of the US TV *Star Trek* space-operas *The Next Generation*, *Deep Space Nine* and *Voyager*. In these, the 'Holodeck' is the 'virtual entertainment' zone of the Starfleet vessels, where crew members can, for relaxation, order up 3-D hologram fantasies, and enter into them, dressing up to take on characters, and actually playing out a form of live dramatic improvisation with holo-projection co-performers, who appear, and act, completely convincingly as fellow human beings, within a consistent, populated, holo-projection dramatic world. This is an electronic near-future vision of the role-playing board-games such as *Dungeons and Dragons* which developed into flourishing Live Role Playing Games and Computer Role Playing Games (such as GraphSim's *Baldur's Gate*) during the 1990s.

In *Star Trek Voyager*, Captain Janeway's mission is to lead the crew of her Starfleet vessel, *Voyager*, which is lost in the unexplored Delta Quadrant of Space, through whatever hardships and adventures they might encounter, in a determined attempt to get home – to Earth Starbase. The ship's doctor is himself a hologram projection, akin to the characters on the Holodeck, but more sophisticated, programmed with an inexhaustible fund of medical knowledge, and the ability to develop increasingly human traits. By *Episode 7.5: Flesh & Blood (Part II)* (David Livingston, USA 2000), the endearing Doctor has long been released from his original confinement in medical quarters by means of mobile holo-projection, so he is able to participate in the life of the ship like any other crew member. In this episode, his loyalties are put to the test: a group of hologram-characters have escaped in a ship fitted with a powerful holo-projector, whose presence allows them to pursue their own lives free from their human creator/captors. Like the Doctor on board *Voyager*, these rebel holograms can move freely, and interact with solid objects (such as spaceship controls), just like human beings. They kidnap the Doctor, beaming him aboard not only for his special skills and talents, but in the spirit of freedom – to save him from his subservience to humans. Tired of the life of continual violence and terror imposed upon them by the humans who programme them to play roles (most often those of ever-revivable, ever-pursuable, and woundable, quarry, or ever-torturable victims) in complex war and shoot'em up fantasies, these rebel holos are determined to find a planet whose atmosphere is inimical to humans, and set up a peaceful colony there for silicon-based life – free from the domination of carbon-based life-forms, which seem to be obsessed with power and pain. The Doctor must decide whether to side with the holos against the humans or vice versa. In fact, he discovers that the holo leading the revolution (programmed originally by humans, of course) suffers from megalomania: he sees himself more and more not only as a liberator, but as a messiah for his people. His idealism is shown to be backed by all the ruthlessness of the most unstoppable computer-game zombie, or human racist totalitarian dictator – and his destructive capacities are directed against the Doctor's human crew-mates, as against all humans. The Doctor, who has sworn the Hippocratic oath, and is in any case personally loyal to the peaceable multispecies aims of the Federation, after arguing in vain with the rebel leader against his policy of exterminating all carbon-based or other life which might stand in the holos' way, reluctantly but consciously makes himself instrumental in turning off the rebels' projector, and so neutralising the rebellion.

In 2000, this vision of holographic characters who look exactly like human beings, and have the artificial intelligence to interact with them in dramatic role-play in physical space, was very far from being achievable anywhere but on a Starfleet vessel studio-set, using human actors to impersonate holograms for TV cameras. 'Interactive computer games', such as the Lara Croft *Tomb Raider* (Core Design / Eidos Interactive) series, which defined the state of the art from Lara's first appearance on the Sony Playstation console in 1996, in 2000 featured pixellating steerable avatar/characters which, under expert direction from players, could execute a wide range of very impressive acrobatic moves (usually with destructive or evasive intent), but bore only a superficial relation to human (or live actor) characters. Gameplay focussed on adrenaline-driven adventure, motivated by the need to solve puzzles and negotiate hostile environments populated by death-dealing enemies, against time. Character and storylines tended to be established through (sometimes very elaborate) pre-gameplay 'backstories', in the form of linear pre-recorded animated sequences, known as 'Full Motion Videos' (FMVs). Play proceeded through stepped levels of difficulty, and at the end of each level there might be (as in the Lara Croft adventures) a sequence of inserted 'cut scenes', setting up the next stage in the gameplay, both the tasks to be achieved and the fictional rationale behind the adventure. The quality of the animation in these essentially narrative sequences was restricted by the technical parameters of computer memory-capacity and processing power as well as screen resolution: the resulting crudeness of many features, such as coarse facial animation, or the inability to use hands to pick up in-game objects convincingly, were a far cry from fully articulated dramatic fantasy-characters, such as those generally found in feature animation from the Disney studios, or animation directors Myazaki, Bluth or Lasseter, leave alone the holo-performers of *Star Trek*. The single-hero epic adventure was the standard game form, since the difficulties of processing multiple characters in real time were insurmountable.

Even in 2002, with the development of faster, more powerful computers and software, sophisticated multiplayer and online games such as *Unreal Tournament* (Epic Games / GT Interactive 1999–) or the later *Baldur's Gate* games (GraphSim 1997–), realtime interactive computer animation, though more refined than in the early 3D-navigable games of the 1990s, does not allow for very complex characterisation or in-game dramatisation. On the whole, multi-character action has to take place in miniature, fielding crude figures and blocky backgrounds, while complex graphics and rich designs have to be reserved for the less frenetic and exciting parts of a story, which are outside the realtime control of players – such as 'cut scenes' or 'FMVs'. Nonetheless, interactive gameplay offers a potential continuum with screen drama in cinema and on TV. With these twentieth-century media, the audience sits in an auditorium or in front of a TV-set, watching the action unfold, engaged by the screen language, delighted by the play of illusion, immersed in imaginative identification with characters and situations, held by the suspense of an unfolding plot. In computer-enabled adventure gameplay, the gamer has a new kind of power – the hero can do nothing unless the gamer uses the controls to operate the animated avatar representing her or him. Swift and sometimes gruesome death will inevitably ensue if the player does not keep the character moving, react rapidly to threat or attack, and take evasive or aggressive action on behalf of the hero. The identification between interactor/player and character avatar is thus very immediate, the avatar standing for the player in a different way from the way actors in screen drama stand for the audience through imaginative empathy – although dramatised avatars like Lara can summon up traces of this kind of empathy. Lara Croft, in *Tomb Raider: Last Chronicles* (CoreDesign / Eidos Interactive 2001), as in all her previous adventure games, introduces herself to the player in an (optional) introductory sequence set in her Elizabethan mansion home, where the player can learn to steer the Lara avatar through a series of gymnastic moves without having to contend with attacks from hidden enemies, and at the same time discover Lara's character and idiosyncrasies through her personal taste in dwelling-place and furniture. She has a large library, and a music room equipped with grand piano, as well as a swimming-pool, training track and gym, and her bedroom and bathroom are luxurious and sensual. A traditional English butler hovers in the background. Via her voice-over, Lara speaks directly to the player, explaining how 'you can make me jump across the gap by pressing the control button marked X'. This intimacy with Lara creates a sense of character-based drama, which the FMV scenarios and cut scenes explaining the story-element of the game reinforce by offering insight into Lara's motives (not financial gain but a spirit of adventure and justice) and the way others regard her (a formidable, inventive, tenacious, but fair, enemy; a courageous, generous, loyal and desirable friend).

Once the play begins, and the avatar, steered by the player, enters the virtual space of the gameworld, the player 'becomes' Lara as she journeys through the screen, and beyond, into fully navigable 3-D cyberspace. This is a very different screenworld from that of traditional film or TV, although when, as in the case of Lara, a game is played on an entertainment console (here, the Sony Playstation, though it could as well be Microsoft's X-Box, or any other), the player is in fact looking at the monitor of the familiar TV set, recognised mediator of character-based drama, into which the console plugs. However, the space on the other side of the glass screen offers more than the cinematic illusion of real space created by traditional camerawork, editing and sound: it is in actual fact fully 3-dimensional. The avatar (steered by the player) can move through it in any direction, while the computer realises the environment in real time as the character (player) requires it. The space, like the narrative, is fully explorable. The illusion of immersive 3-dimensionality created in the twentieth century in cinema, by projecting light onto a stretched screen and activating imaginative power in the audience through powerful dramatic structures, convincing characterisation, and the conventions of effective audiovisual screen language, is transposed in the twenty-first century into navigable expressive space which the player enters in the form of an avatar, who has no life unless the player imparts it. The avatar figure of Lara (or any other adventure game hero) will literally stand still until shot down by predators if the player does not steer it. All the power the formidably competent Lara deploys depends on the player activating and sustaining her life. The player becomes responsible not only for Lara's attaining her quest(s) – to locate the talisman(s) and return home – but also for Lara's very survival. This is a powerful form of identification, fuelled by a strong adrenaline reaction, since the player, both in the persona of Lara and in the actual gameplay, is constantly challenged to react fast to attack, eyes and ears fully alert, manual dexterity under high pressure to push the right buttons on the control pad at the right time, simultaneously remembering and navigating the layout of the gameworld, while trying to predict the moves of the enemy and solve Lara's increasingly challenging puzzle-problems, in order to get her safely to the end of the adventure, and home.

In such 3-D computer games, the drama itself, the story, is far less prominent than in the narrative of film and TV. In games, narrativity is about exploration and discovery, about challenge and successful negotiation, more than about plot or the aesthetic pleasures of immersive illusion in the life of another. But it also offers a new narrative stance, one where the virtual camera is not primarily the director's instrument of mediation, whereby audience reaction, comprehension of the narrative and attitude to the characters can be controlled, but a tool to aid the player in navigating the environment and seeing and acting on behalf of Lara. It incorporates camera angles and moves which violate every principle of cinematic realism, where, as long as the camera is placed in a position natural to the human eye (or, in the case of ET, the Extra-Terrestrial's eye), the convention of movie verisimilitude is so powerful that the audience tends to accept easily the truth of the illusory screen world. In real-time 3-D virtual environments (RT3DVE), the 'eye' of the camera can be programmed to simulate movement anywhere in relation to the characters and the 'location', since the camera itself has no physical embodiment. In this sense, the computer-game screen is more like the canvas of a painter than the traditional cinema screen, since every image, including the human figures, is designed by visual artists and designers. In theory, it shares the freedom which stories made in 2-D character-based animation have traditionally exploited, showcased in Disney's *Fantasia* (Ben Sharpsteen, James Algar, Joe Grant, Dick Heumer, USA 1940; compare *Fantasia 2000* (Eric Goldberg, Hendel Butoy, Pixote Hunt, James Algar, Francis Glebas, Gaetan Brizzi and Paul Brizzi, Don Hahn, USA 2000)): the ability to create abstract settings and improbable action – epitomised by Warner Brothers' 'stretch and squeeze' techniques, where Loony Tunes characters can be battered into blots, shattered into fragments or spattered into dots, and still spring back into shape, full of vitality – and the possibility of showing the action from any angle the artist desires. The properties of the screen world in RT3DVE have the potential to be very different from those of the 'real' world, or even of the studio set. They seem to reach farther towards the cinema's traditional desire to release the world beyond the proscenium arch and open it to the audience – dissolve the threshold, allowing admittance through the 2-D screen into 3-dimensional, illusory, existence . . . somewhere over the Rainbow.

However, in RT3DVE as conceived at the beginning of the twenty-first century, it is technically hard to stage even epic/heroic stories, where the characters and drama are centred round a single hero, in part precisely because of the lack of restriction in the rhetoric of mediation. If the camera is totally free in space, how will it mediate the drama in the most striking and effective way? If players determine where and when characters perform their action, how can suspense, rhythm and satisfying imaginative immersion in a story world be achieved, and overall sense made of the narrative? On the whole, few have attempted to meet these narrative challenges, and as, with the refinement of technology, computer games emerge as an increasingly powerful form, there has been a strong tendency to focus research and development on the properties unique to the medium, rather than those it shares with more traditional art forms, including theatre, cinema and TV. At the beginning of 2002, the target audience for interactive narrative game titles had scarcely extended beyond the group of males aged 10–25, who formed the earliest buying-sector for computer games. Nonetheless, some Japanese products – such as the Squaresoft *Final Fantasy* series – with more narrative elements than dominant US or EU titles, have achieved popularity and a devoted following in the West as well as the East, which suggests that the little-explored potential of narrative in RT3DVE may yet develop fruitfully. Once critique of the field has progressed beyond opposing the 'ludic' qualities of games to their 'narrative' qualities, insisting that these must inevitably clash – with the inference that narrative must detract from gameplay (proven to be popular and lucrative) and should not therefore be developed in the commercial context – interesting movement in this area may start to take place outside the non-commercial zones where experiment has largely been confined hitherto.

Whatever the limitations or ambitions of computer and console games, their aesthetics and conventions have embedded themselves in the language of 21st-century cinema. Alongside her phenomenal popularity as a game hero, the characterisation behind the Lara Croft figure, and the classic (Professor Challenger / Captain Marvel / Indiana Jones) structure of her adventures, encouraged film-director Simon West to make *Tomb Raider* the movie (UK 2001). The high-budget film opens with a sequence showing Lara battling a nasty and very big techno-monster, in a set very similar to the Egyptian and other ancient temple and tomb sites which are the typical environment of her 3-D animated adventures. This physical set is of the genre found in movies like *Indiana Jones and the Last Crusade* (Steven Spielberg, USA 1989, designed by Elliot Scott), or *The Fifth Element* (Luc Besson, France 1997, designed by Dan Weil). However, the stunts Lara (played by Angelina Jolie) is required to perform cannot be achieved by a human actor without the help of pulleys and wires, and the opening monster-fight, like the even more spectacular scene of Lara's reaction to an invasion of her home, uses digital compositing both to remove the harnesses and strings, and to permit angles of view which are reminiscent of the virtual camerawork developed in RT3DVE, with which gameplayers of the *Tomb Raider* series are familiar. Cinematic realism is abandoned in favour of reproducing on the big screen – that traditional locus of magnificent illusion – using a human performer, the impossible performance of the animated Lara-avatar in computer gamespace, all viewed as though through the free eye of the virtual camera – though for the cinema the effects had in fact to be achieved using 35 mm film-cameras supplemented with digitally post-produced special effects.

The impact of this feedback of gamespace into cinematic space on the narrative structure of the movie is intriguing. West's approach is to use game language, and some game conventions – Lara in the movie, like her virtual original, is still out questing for a talisman in exotic locations featuring an underground temple/tomb (designed by Kirk Petruccelli), and she still has to pull switches concealed in improbable places to open secret passages, and perform extraordinary feats of physical daring to attain her ends. But West chooses to re-characterise Lara in the film as a sultry babe, wafting in her free moments through veiled boudoirs, who takes showers in semi-transparent stalls before and after her exertions (watched lasciviously by the camera), and is obsessed not so much with the spirit of adventure and justice, as by the need to recover contact with her absent father (also responsible for her present adventure). Lara in the game is self-sufficient as well as self-reliant; in the movie, she depends on her young (male) techy nerd inventor much as Bond depends on Q. In the game, Lara's adventures do not involve romance; in the movie, Alex West (not to be confused with director Simon West), almost as good at tomb-raiding and stunts as she is herself, rivals Lara, and flirts with her. In the end, Lara reveals her feminine susceptibility to his charms by risking all to save his life – and by her bold gesture also achieves final approval from her dead father. Lara attains both these ends through the daring manipulation of the time-tokens which are the object of her quest, and which allow her to control time long enough to prevent Alex West's wound being fatal, and to pay a visit to her father before his death. Home at last, movie Lara Croft is finally able to abandon her (very brief) combat clothing in favour of a thin ('feminine') frock, to pay her last respects to the paternal memorial at the heart of her mansion. Lara has been on a long and tiring adventure – but at last, she has come home to Daddy. Hollywood conventions for the design and delivery of screen heroines are very powerful – movie Lara is quite unlike the unprecedentedly popular, free-spirited, heroine of the *Tomb Raider* games, who quests ever on through open-ended adventure frameworks, as much her own mistress in each episode as her predecessors Captain Marvel, Professor Challenger, Sherlock Holmes or Indiana Jones. Game-Lara seems to be a figure with whom gameplayers of both genders find it easy to identify and play, to judge by the sales of the games.

The *Tomb Raider* movie was, perhaps partly because her fans were disappointed in Lara's cinematic character, not a commercial or critical success, despite its large budget, splendid design and stunning special effects – plus a central performance by Angelina Jolie (whose real-life father, actor Jon Voight, plays Lara's father in the movie) which succeeds surprisingly well in delivering a moderately believable Lara, despite the conflicts between the original virtual character and her movie counterpart, and all the screentime spent strapped in harness suspended at improbable angles under monstrous cameras, sweating out the close-up parts of the Lara stunts – not to mention the camera's determined pursuit of Jolie's swelling bosom beneath the clinging lycra of her sleek new Lara costume. The amalgam of virtual camera-strategy and game convention with traditional cinematic language and story structure has not quite achieved the balance needed to make it function effectively as narrative. Interestingly, in 2002, Lara Croft in her new RT3DVE game adventure, *The Angel of Darkness*, looks strikingly like Angelina Jolie, who was herself originally cast partly for a physical resemblance to virtual Lara – and her costume has been modified to look more like the film version than the earlier game version – game into film, film into game.

As the twenty-first century progresses, and digital media expand, overlap and become standardised, the two-way influence between movies and computer games is set to continue to develop and modify screen language and narrative stance, as well as structure, in both. The movie *Final Fantasy* (Hironobu Sakaguchi, Japan 2001) fields a cast of motion-captured animated performers rather than live actors, so its look is identical to that of the high-resolution RT3D virtual world and its avatar denizens, delivered by the 2002-generation of fast computers and entertainment consoles; Aki Ross' quest to pick up all the elements she needs to defeat the alien monsters, so that she and her companions can return home to Earth, can be seen as a gamelike space-version of Dorothy's adventures in the Land of Oz. However, the delight of the movie animators in showing off the illusions they can achieve, and the time it takes to demonstrate lighting effects, or stage tosses of the head which reveal the lifelike movement of computer-animated hair – a triumph of computer animation, but of little interest to audiences in search of immersive storytelling – slow down the pace of the action and detract from dramatic engagement. To engage a cinema audience, computer-, or any other kind of, animated characters need not only a gripping plot, but also aesthetically appealing environments with plenty of variety, good dialogue, good voice performance, interesting character visualisation, choreography and mise-en-scène, plus camera and editing strategies which capture and hold attention and fire the imagination. If they are to exploit the power of screen narrative for animated characters, which has been developed to a highly sophisticated level by Disney, Warner Bros and Manga, writers and directors evidently need different skills from the ones computer games designers coming out of software design and computer graphics backgrounds usually have.

The genre where games and movies have cross-fertilised each other most successfully seems to be the Shaolin/Kung-Fu arena, perhaps for the obvious reason that since its *floreat* in the 1970s, the Taiwan-Kung-Fu film has depended for its success on the astonishing physical prowess of its stars, on its special-effects-aided stunts, contrived and breathtaking choreography, fast-moving plots and strongly delineated traditional character-types. Ang Lee's internationally successful big-screen movie, *Crouching Tiger Hidden Dragon* (China 2001), whose melodramatic and romantic plot shows how a young, untrained and misled martial arts practitioner learns about the spiritual and ethical aspects of her chosen way from an older expert (Michelle Yeogh), whose tragic love-story forms the backdrop to the action, uses many of the camera angles, settings and conventions of popular martial-arts combat computer-games – such as Playstation's *Tekken* (Namco/SCEE, 1996–).

The *Tekken* series, which can be played by two players in competition, like traditional games, or by one player against the computer, has been enduringly popular. Because the moves of martial arts are precisely choreographed, motion-captured performances (where human martial-arts experts wearing sensors which are tracked digitally provide recordings of the basic movements from which the animated figures, like the Lara Croft game-figure, are generated) can be combined and recombined instantly by the computer according to set rules, in response to the actions of players. Each character in the cast of fighters has specific skills and talents, and players can match one against the other in many configurations. Players acquire increased manual dexterity and hand–eye co-ordination, skill in understanding the principles of combat and the form of each participant, and hone their own synchronisation and identification with the in-game avatars, with each new game they play. Steering/inhabiting these character-avatars, even the most sluggish couch potato can achieve the feats of a Michelle Yeogh or a Bruce Lee in the virtual world. In addition, each fighter has a backstory, which the curious can explore within the game-world, and which may aid empathic identification. Some of these characters are dramatised in some detail in the spin-off animated movie, *Tekken the Motion Picture* (Kunihisa Sugishima, Japan 1997), which tells a traditional Japanese Manga-animation-type martial-arts hero-tale, of how a brother and sister grow up through a series of trials to support each other against the rest of the world, and defy their father's scorn, using the strengths they develop. In *Tekken*, there is no conflict between the animated genre and the game genre – each takes on a different aspect of the characters' world, the movie concentrating on narrative drama, while the game focusses on combat skills. Together, they build a fiction-world which can be expanded both with new movie episodes and with new game characters. The animation style in movie and game share some features (though one is in 2-D and the other in 3-D), but on the whole the animated movie world and the game world appear to be regarded as distinct zones, complementing each other and dealing with the same characters, but not in the same medium – just as tales of Robin Hood and his Merry Men exist simultaneously as told stories, books, live-action movies, TV series and animated 2-D cartoons.

Generic movies have provided the most inspiration for in-game narrative in computer games, and, in 2002, PlayStation 2's *Grand Theft Auto III* (Rockstar / DMA Design), clearly using the conventions of the American gangster movie, offers cinematic cut scenes and spoken dialogue much closer to movie dialogue, including impressive voice performances, than anything in previous adventure games. The gamer has to participate in a world of organised crime, understanding, interpreting and negotiating it via gangster movie convention. This movement towards fictional-genre role-playing as part of action-based computer games, like that inspired by Tolkien's work in the fantasy-arena, opens up a promising area of RT3DVE narrative potential, closer to theatrical or musical improvisation than traditional film or TV narrative. Nonetheless, as long as the screen is the locus of its performance, however collaborative and improvised, audiovisual screen language will remain fundamental to the expression of these evolving genres.

From another perspective, the 'haunted house' or 'house of horrors' movie genre was adapted for Playstation's *Resident Evil* series (Capcom / Virgin Interactive Entertainment / 3DO /Eidos Interactive, 1996–), with most success where it uses horror-movie tricks (clearly inspired by those of film-director John

Carpenter (e.g. in *Hallowee'en*, USA 1978) and his peers) to build moments of shock within the game, as the player-avatar is repeatedly attacked from unexpected angles by apparently indestructible monsters, usually just as the hero believes they have finally been destroyed, and the territory seems safe. The *Silent Hill* dramatic adventure games (Konami/Konami 1999–), some of the first to experiment with sophisticated motion-capture-based 3-D character animation, devised an intriguing synthesis of the ethos of the movie genre typified by Wolf Rilla's *Village of the Damned* (UK 1960) with computer quest-story, elegantly exploiting the potential of participative interactivity in an eerie expressionist American suburbscape. *Silent Hill* creates a building sense of malice through emotive virtual sets and audio, echoing the misty melancholy rife with hidden threat beloved of sixties UK Hammer movie-designers such as Bernard Robinson, immersing the interactor in a subjective reality where the quest for a helpless young missing daughter leads the avatar-character through bewildering and chilling, mocking illusions, satanic deceptions and blind alleys. A great strength of *Silent Hill* is the understanding and discretion with which it transposes cinematic narrative techniques to the virtual environment: unlike those games which merely copy a cinematic scene – or lift moments from one (not always from appropriate movies, rarely as part of a coherent dramatic aesthetic) – into their gameworlds, *Silent Hill* adopts the subjective narrative stance of the horror-movie genre to lure the player into a world of uncertainty and deceit, which increase the deeper you go. The spatial and visual design of the game actively express the malevolent entity which is toying with the player/hero, who is thus subjectively affected by the conditions that are undermining and coercing the avatar-protagonist in the game. Imaginative identification is cumulative, and effective.

Building on the slender narrative tradition emerging in computer games in the 1990s, the PlayStation martial-arts adventure *Tenchu Stealth Assassins* (Acquire/Activision 1998) offers a combination of quest, drama and skills in the virtual territory of Ang Lee's big-screen movie, *Crouching Tiger Hidden Dragon*. In *Tenchu*, success is achieved by skill and the ability to act unseen, to outwit your enemies, and in hand-to-hand combat to defeat them through inner focus and perfect mind/body co-ordination, rather than by using brute force or ballistic weaponry (as in most Western shooting-games, or 'Shooters'). Animated 'cut scenes', containing both dramatic dialogue and action, occur relatively frequently in *Tenchu Stealth Assassins*, knitting the story and the gameworld together. Players traverse territory familiar from the Kung-Fu movie-landscape, featuring sloping dragon-crested roofs above shadow-filled enclosed courtyards, paper screens, fantastic leaps, and astoundingly accurate, perfectly timed and sure-footed cat-like drops. The cinematic screen illusion is not, however, dominant – dialogue, as well as being spoken, is printed as subtitles on the screen, and the voice-script and performance have none of the panache of *Grand Theft Auto III* (written by Dan Houser, Paul Kurowski and James Worrall) – with the result that the net effect in *Tenchu* is closer to a kind of audio-animated comic book, or graphic novel, than to traditional movie audiovisual language. The imaginative effort required to really identify with the characters, despite the use of some dramatic camera-strategies and filmic conventions, and the staging of dramatic scenes, is considerable. However, it is an effort young imaginations, aided by the adrenaline drive of the combats and the challenge of achieving martial-arts stealth-skills and control, are usually strong enough to make without difficulty, and this genre has obvious potential for narrative development in the Ang Lee direction.

This brief review of screen adventure from 1922 to 2002 suggests that the power of screen narrative does not stem only from strong story-patterns and recognisable generic frameworks, such as the traditional epic *Wizard of Oz* dream/fantasy, where lost and vulnerable heroes, plucked from their normal environment, have to gain skills and confidence through trials which finally enable them to find their way home – matured, improved, transformed and/or wiser for their adventures. Like that of any other narrative form, the power of screen narrative derives substantially from its rhetoric, and the effective use of the expressive conventions of its medium. Over a period of some hundred years, moving-image storytelling has haunted and inhabited a number of screens, from the cinematic – that fragile 2-D membrane for realising shared projected illusions – through the small, brightly illuminated box of the domestic television, which permitted audiences to join the crew of the starship *Voyager*, as it searched for the way back to Earth; to the unflickering computer screen, domain of player-driven avatar action in insubstantial navigable 3-D electronic environments, where Lara Croft eternally seeks her talismans and the ways out of her scenarios. The magic of immersive fictive worlds on all these screens is mediated to the audience and interactor through narrative stance and audio-visual language, expressed by mise-en-scène, cinematic strategies and editing. This will continue to be the case, as, in the electronic era, the content delivered via the public screens of e-cinema and d-cinema and the domestic interactive digital TV screen / entertainment centre merge more and more.

Movies made for the early 2-D silver screen elicit imaginative identification with action and characters through a sense of wonder and delight in the illusion, which transports us into the realm of fantasy; late twentieth-century movies add techniques of cinematic realism, sophisticated stereo sound and visual special effects to draw us, through 'sensory illusion' and camera-steered empathetic identification with characters, farther into the world beyond the proscenium – even extending the illusion of that world out into the auditorium to envelop us. The twenty-first century brings the computer as a creative medium, a tool and a delivery mechanism into the foreground, developing ephemeral real-time-generated 3-D navigable expressive environments and steerable avatar characters, which enable interactors to traverse fictive storyspace in the persona of a superactive participant adventurer.

Each age develops its own technologies, and inherits traditions; and these interact with storyforms and the human need for, and delight in, storymaking, make-believe, illusion and the joys of the creative imagination, to produce a range of narrative shapes and patterns. There are well-told tales and badly told tales in all media. In an era where digital audiovisual media operate alongside pen and paper, or replace them, an analysis of the audiovisual means of expression in 2-D and 3-D screen environments offers important insights into how the telling embodies the told. The tales themselves are devised, assembled and passed on by the tellers, who use the media available to them, but, in the end, the power of narrative, whether in the age of Portia and Miranda, of Dorothy, or of Lara, depends not on content or form alone: but on the effective combination of matter with manner, narrative with narration – the expressive as well as the structural aspects of the fusion that is medium and message.

Film quickly developed camera and sound techniques to situate audiences within the narrative frame of a movie, early adopting the subjective viewpoint of hero-centred storytelling and adapting it for the cinema, allowing audiences to travel through the screen to dreamland. Audiovisual language works directly through the sense of vision and the sense of hearing, allowing effortless audience immersion in a fictive world, and for a hundred years it has been refined to enable directors to shift an audience's viewpoint and sense of identity to adopt the persona, and share the experience, of a Dorothy, a Paul Hackett, a Neo, a Lara Croft or an ET. Spectators can learn with Dorothy that loyalty, a pure heart and the desire to do good will develop the courage and intelligence to overcome the most threatening evil, and bring them safe home again; with Paul audiences can face unconscious fears and dark anxieties, and be reborn after their night terrors into the real world at the light of day; with Neo, they can defeat, in cyberspace, the powers that threaten to drain us of our life-forces, and emerge reassured that such victory is possible; or they can empathise with the outsider, in the form of a gnarly little alien, and, through ET's eyes, view the normative values of their own society from a new perspective. On the movie screen, spectators can watch the dauntless Lara Croft accomplish impossible feats, and imagine themselves turning back time to seek reconciliation with a lost father, or save a worthy and attractive antagonist. In 3-D virtual reality, players can actually imbue Lara with their own life-energy and skills, and, in her persona, pursue endless adventure through exotic locations and dream landscapes, overcoming the forces of evil, and recovering the talismans that will save the world. In the twenty-first century, the power of screen narrative fuses representation, imagination, empathy, sensual engagement and physical participation to transport us to new worlds and provide us with new experiences, to feed and lead our imaginations, and to shift our viewpoints, redefining 'them' and 'us' – a mighty power indeed . . .

FURTHER READING

Reiser, M. and A. Zapp, eds. (2002). *New Screen Media: Cinema/Art/Narrative*. London: British Film Institute.

Ryan, Marie Laure (2001). *Narrative as Virtual Reality: Immersion and Interactivity in Literature and Electronic Media*. Baltimore: Johns Hopkins University Press.

Ryan, Marie Laure, ed. (2004). *Narrative Across Media: The Languages of Storytelling*. Lincoln: University of Nebraska Press.

5 The power of death in life

ELISABETH BRONFEN

Between the solitary and the social

Death is a solitary, individual and incommunicable event, perhaps the most private and intimate moment in the cycle of human life. Whether it marks, in religious terms, an exchange – whereby the dissolution of the body is contiguous with an entry into a new spiritual existence and, thus, the return to divinity – or whether, in the more secular encoding of what Sigmund Freud calls 'the death drive', it merely initiates the return to that tensionless, undifferentiated state of the inanimate that is beyond, grounding and prefiguring biological and social human existence, in either case the finality of death is generally acknowledged as the one certainty in any given life. It is the powerful fact against which, and in relation to which, all mortal existence is measured. At the same time it is impossible to know in advance what the experience of dying will be like, as it is also impossible to transmit any precise and definitive knowledge of this event to those who survive the death of another. In that sense death is also the powerful limit of all mortal knowledge; its ground and its vanishing point.

Yet dying, burial and commemoration are always also public matters. As cultural anthropology has shown, death, in that it removes a social being from society, is conceived as a wound to the community at large and a threatening signal of its own impermanence. The dying person, and then the corpse of the deceased, occupy a liminal[1] place, no longer fully present in the world of the living and about to pass into a state inaccessible to them. Rituals of mourning, falling into two phases, serve to redress the disempowering cut that the loss of a group member entails, creating a new identity for the deceased and

---

[1] *Liminal*: from the Latin word *limen*, meaning *threshold*.

Power, edited by Alan Blackwell and David MacKay. Published by Cambridge University Press.
© Darwin College 2005.

reintegrating her or him back into the community of the survivors. On the one hand, a phase of disintegration marks the dangerous period of temporal disposal of the corpse and the mourners' separation from everyday life, celebrating loss, vulnerability and fallibility. On the other hand, a phase of reinstallation or second burial reasserts society because it emerges triumphant over death. Thus rituals of mourning, acknowledging the wound to the living that death entails, always also work with the assumption that death is a regeneration of life. In particular, the conceptual translation of death into sacrifice serves as a cultural ruse that works against death. The sacrificial victim, representing the community at large, but placed in the position of liminality between the living and the dead, draws all the evil or pollution of death onto its body. Its expulsion is then, in turn, contiguous with purifying the community of the living from death. While the loss of a cherished family or community member evokes grief and the pain of loss for the survivors, viewing and commemorating the death of another is also a moment of power and triumph. Horror and distress at the sight of death turn into satisfaction since the survivors are not themselves dead. Visual or narrative representations of death, meant to comfort and reassure the bereaved survivors, as is the case in tragic drama and elegiac poetry, ultimately serve to negotiate a given culture's attitudes to survival. Signalling such a gesture of recovery after the disempowering impact of loss, a given society will perpetuate stories about sacrifice, execution, martyrdom and commemoration so as to affirm its belief in retribution, resurrection or salvation, much as an individual family will generate stories about its deceased ancestry to express its coherence after the loss of one of its members.

For this reason it is one of the great plot conventions to use the funeral statue of a deceased as the catalyst for a tale about his or her symbolic reinstallation within the community of the survivors by virtue of the commemorative narrative this calls forth. For, within the funeral ritual, the actual corpse has been removed from its community and replaced by a piece of sculpture resembling it. At the same time, this uncanny doubling of decomposing body and inanimate body elicits a second type of representation – the tale the survivor has to tell. As a particularly salient example for the exchange between life and death that is publicly performed by virtue of such funerary representation one might take Joseph L. Mankiewick's film *The Barefoot Contessa* (1954). The film significantly begins at a cemetery, where the friends as well as the fans of the late Hollywood star Maria Vargas (Ava Gardner) have come together one rainy morning to take part in her burial. As the camera moves from an establishing shot, showing the crowd gathered in front of a marble statue of the deceased, to a close shot of the director Dawes (Humphrey Bogart), who had initially discovered the young woman while she was still dancing in a bar in Spain, his voice-over begins. We hear him recall how he had first met her in a bar in Madrid, convinced her to return with him to Hollywood, and directed her in the films that came to make her international fame. In the course of *The Barefoot Contessa*, Mankiewick shifts between several narrators, moving from the director to the husband of the deceased as well as to others who knew her, so that over the dead body of the Hollywood glamour icon each of the survivors is able to weave the story that will let him go on living, precisely by explaining his relation to Maria Vargas, and to a certain degree thus also explaining the implications her death has for his own survival. To support this shift in narrative perspective, Mankiewick returns to the establishing scene at the cemetery after Dawes has finished his part of the tale, as though to emphasise not only that Maria Vargas' death functions as the catalyst for all the narratives that commemorate her, but also that, like the funerary statue standing in for the dead body buried beneath the grave plate, these tales help the mourners to reinstall the dead woman into the symbolic community of the living precisely by turning her into a sign, namely a narrative they can share with others. It is, thus, also significant that her death – she is shot by her jealous husband, the Conte Vicenzo Torlato Favrini – occurs after the funerary statue, which we see at the very beginning rising high above the heads of the mourners during the funeral ceremony, has been completed, as though this aesthetic representation were already the mark of a death *avant la lettre*.

In discussing the more personal aspects of grieving the loss of a beloved person, Freud has suggested that the normal affect of mourning bears resemblance to melancholia. In both cases the response to the loss of a loved one is a turning away from all worldly activity such that the mourner instead clings almost exclusively to the deceased love object. However, whereas melancholia describes a pathological condition that arises because the afflicted person is unwilling to give up his or her libidinal investment in the lost love object, in the case of mourning, the lost love object is ultimately decathected[2], but only

---

[2] *Decathected*: psychoanalytic term meaning the removal of psychic energy from a specific goal. From *cathexis*, a concentration of psychic energy.

after an extended period during which the survivor works through the memories, expectations and affects attached to the dead. In this sense the type of narrative commemoration cinematically performed by Mankiewick functions analogously to the liminal period, in the course of which remembering the dead allows those who knew Maria Vargas to work through their libidinal investment in her, so that at the end of each of their stories a disinvestment of sorts has been accomplished. Because, as Freud insists, with worldly reality once more gaining the upper hand, the process of mourning comes to an end and the afflicted subject is again liberated from the painful unpleasure that was cultivated during the mourning process. The narratives told in the course of *The Barefoot Contessa* thus structurally double the exchange between corpse and statue, in that they, too, allow the survivors to draw a clear boundary between themselves and the deceased precisely by exchanging her bodily presence into an absence, referred to by a narrative commemorative text. Within the larger context of memorial practices, rituals such as attending wakes and séances were designed as further ways meant to assist such a working-through process, for they allow the mourner to enter into a dialogue with the deceased, but under the condition that this exchange will ultimately find closure, in the first case when the body is buried, or, in the latter, when the spirit is once again released. Visits to cemeteries, or in the case of those who died as a result of wars and other political catastrophes, to memorial sites of collective commemoration, furthermore, work with the presupposition that the living no longer harbour a libidinal investment in the lost love objects, even while they are meant to assist the survivors in preserving their memory of the dead. Therein also lies the power of aesthetic representations, revolving around incidents of death; they preserve a recollection of the dead, indeed function as a conversation with the dead, even while ensuring that, at the end of the aesthetic experience, closure is put onto this uncanny exchange.

Historicizing death

Any discussion of the aesthetic rendition of death is thus fraught with contradictions. On the one hand, it must account for the fact that dying is always a solitary act, a highly ambivalent split both for the person dying and for the survivors. It can elicit both psychic distress and serenity, induce a sense of burden and relief and fulfil both a desire for and an anxiety about ending, so that any images or narratives of mortality inevitably touch emotional registers in relation to an event of loss that enmesh the terrifying with the uplifting as well as with the inevitable. What emerges is a highly complex interplay of grief, anger, despair, acceptance and commemoration of the deceased; an interplay so highly personal, individual and specific that it is seemingly performed outside historical and social codes. Indeed, because the transitory nature of human existence and the possibility of an afterlife have always preoccupied the living, because all earthly life is directed towards death and one's conduct is fashioned in view of death and the possibility of salvation, representations of death seem to be an anthropological constant that refuses to be situated historically.

On the other hand, precisely because burial rites are used to reinforce social and political ideas, with tombs and funerary sculptures endorsing concepts of continuity, legitimacy and status, historians have also been eager to demonstrate that different periods are characterised by different cultural images of death and attitudes to it. The most prominent, Philippe Ariès, offers a linear development that begins with an early European acceptance of death as an inevitable fact of life, as an organic and integral part of a harmonious reciprocity between living and death. With the emergence of individualism, however, the destiny of each individual or family takes precedence over that of the community and a new emphasis is placed on the funeral as a sign of social status and material wealth. At the same time the focus on the self provokes a passionate attachment to an existence in the material world and hence a resentment of death. By the mid eighteenth century, for Ariès, an attitude of denial, which links the fear of death to a fascination for it, becomes the norm. While cemeteries are symbolically removed to the outskirts of the city, the dying person and the corpse become objects of erotic, mystic and aesthetic interest. Ariès calls this the 'period of beautiful death', and, in a sense that has permeated well into the late twentieth century, aestheticisation hides the physical signs of mortality and decay so as to mitigate the wound that death inflicts on the survivors. In explicit reference to the iconographic domain of Ophelia and her many visual refigurations serving as an example *par excellence* for such a beautification of death, Charles Laughton stages the beautiful corpse of the young mother Willa Harper (Shelly Winter) in *The Night of the Hunter* (1955), killed by Revd Harry Powell (Robert Mitchum). The psychotic self-styled preacher had discovered from her husband, who had been his cell-mate in prison, that the latter had hidden the money from a bank robbery somewhere in his home, and has now decided to infiltrate the Harper household so as to capture this loot, even if this requires the destruction of the entire family. After showing us the murder, Laughton cuts to a seemingly idyllic scene. A fisherman is sitting in his boat, singing to himself as he is waiting for a fish to take his bait, and then bending over the edge of his boat, once his hook seems to have taken hold of something. Significant about the mise-en-scène Laughton has chosen is that Willa Harper seems to be floating in a liminal zone between life and death. For what the fisherman finds beneath the water's surface is not the sought-for fish, but rather a woman, sitting in an open car, dressed in a white night-gown, her long blond hair waving about her. She is suspended between life and death, not yet decomposed but also no longer of the living. She is, above all, a body on display, as though the car she is sitting in were her frame, the water around her a liquid of preservation. As such, she seamlessly turns from an actual figure on the diegetic level of the film into an aestheticized object, notably that of the fisherman's astonished and transfixed gaze. Because Laughton offers us her image initially through the fisherman's perspective, only to shift to a close shot of her, taken as though from inside the water she has been submerged in, Willa Harper readily transforms for us, the viewers, into an eerie body that is no longer located in any realistic space, and has instead – by virtue of this staging of her dead body – been transferred to a zone of aesthetic refiguration. The violence of her death has been mitigated. She is no longer the victim of a devious madman but rather a figure of timeless beauty, arrested by death but also preserved by the cinematic image.

Yet, there is a seminal contradiction inscribed in strategies of beautifying death aesthetically. Whether through spiritualism, which offers a male-centred domestication of heaven as a continuation or repetition of earthly existence, or through a cultivation of burial and mourning insignia – consolatory literature, elaborate tombstones and pompous cemetery monuments – aesthetic beautification renders the terror and ugliness of death's reality palatable by placing it within the realm of the familiar as well as the imaginary. By the mid nineteenth century, visits to morgues, houses of mourning and wax museums had become comparable to visiting a picture gallery. This death so lavishly represented was, however, no longer death but rather an illusion of art. Yet a seminal contradiction came to be inscribed in this allegedly modern attitude towards death, persisting today in our visual, narrative, cinematic and cyber-representations of violence, war and destruction as well as in the sentimental stories about victims our cultural discourses engender so as to idealise and make into heroes those smitten by death. The more Western culture refuses death the more it imagines and speaks of it. Aestheticisation, meant to hide death, always also articulates mortality, affirming the inevitability of death in the very act of its denial. With death's presence relegated to the margins of the social world, representations of death also turn away from any reference to social reality, only to implant themselves firmly in the register of the imaginary. Reflexivity comes to be inscribed in images of death in that, because their objects of reference are indeterminate, they signify 'as well', 'besides', and 'other'.

Locating at the end of the eighteenth century the epistemic shift that rein-stalls a discourse of mortality, which insists that all knowledge is possible only on the basis of death, Michel Foucault has highlighted the contradiction at issue. Death, which is the absolute measure of life and opens onto the truth of human existence, is also that event which life, in daily practice, must resist. The metaphor Foucault uses to illustrate how death is the limit and centre toward and against which all strategies of self-representation are directed, is that of a mirror to infinity erected vertically against death: 'Headed toward death, language turns back upon itself. To stop this death which would stop it, it possesses but a single power: that of giving birth to its own image in a play of mirrors that has no limit' (1977: 54). As death becomes the privileged cipher for heroic, sentimental, erotic and horrific stories about the survival and continuity of culture, about the possibility and limits of its knowledge, it self-consciously implements the affinity between mortality and the endless reduplication of language. What is called forth is a literature where aesthetic language is self-consciously made into a trope[3] that refers to itself, seeking to transgress the limit posed by death, even as it is nourished by the radical impossibility of fully encompassing this alterity. In a similar manner Martin Heidegger has argued that all life is a 'being toward death', with all existence forcing the human subject into recognition of this abyss, into a realisation that one is never at home in the world. Such an encounter with the nothingness of the veiledness (*Verhülltheit*) of death, although it initially calls forth anxiety, ultimately leads to the recognition of the truth of being, namely, an experience of the ontological difference between being (*Sein*) and beingness (*Seiendes*), with the former overcoming the latter. Representation for Heidegger is authentic when it bows into the silence evoked by the measurelessness of death, while any language that avoids death is for

---

[3] *Trope, tropic*: in literary theory, a rhetorical or figurative device.

him mere idle chatter. Similarly, Georges Bataille describes the trajectory of human existence as a move from a discontinuous state of earthly fracture and difference to a state of unlimited continuity through death.

Speaking of the aesthetic rendition of death thus ultimately brings into play the question of power residing in misrepresentation, for the paradox inherent in representations of death is that this 'death' is always culturally constructed and performed within a given historically specific philosophical and anthropological discourse on mortality, resurrection and immortality. Since death lies outside any living subject's personal or collective realm of experience, this 'death', which is always already representation, can only be rendered as an idea, not something known as a bodily sensation. This idea, furthermore, involves imagery not directly belonging to it, so that it is always figural, and the privileged trope for other values to boot. Placed beyond the register of what the living subject can know, 'death' can only be read as a signifier with an incessantly receding, ungraspable signified, invariably always pointing back reflexively to other signifiers[4]. Death remains outside clear categories. It is nowhere, because it is only a gap, a cut, a transition between the living body and the corpse, a before (the painful fear, the serene joy of the dying person) and an after (the mourning of the survivor); which is to say, an ungraspable point, lacking any empirical object. The final images of Ang Lee's *Crouching Tiger Hidden Dragon* forcefully illustrate the power of the state of suspension that is at stake, given that figurations of death necessarily oscillate between an ungraspable point of reference in experience and signs that reflexively point back to themselves. The young noblewoman Jen Yu, whose recklessness has made her responsible for the death of the master warrior (Chow Yun Fat), has come to a monastery, high up in the mountains, with her forbidden lover, the robber prince Lo Dark Cloud (Chen Chang). Because she is at an impasse in her life, unwilling to marry the nobleman her father has designated for her, unable to simply run off with the robber prince she loves, but also guilty because of the death she has caused, she decides to perform this psychic and social border-situation bodily. She turns to Lo and asks how he wishes their story to end, and, after he has assured her that all he wants is for them to return to the desert, where they were once so happy together, she jumps off the mountain. In so doing, she has

---

[4] *Signifier*: in Saussurean linguistics, the *signifier* is the written or spoken word, arbitrarily chosen, which represents the *signified*, or concept referred to. The letters 'c-a-t' refer to, but have no natural link to, the actual animal.

recourse to a legend claiming that, if one's belief is strong, miracles can happen. Yet significant about Ang Lee's mise-en-scène is the fact that the final image of Jen Yu is fundamentally ambivalent. All we see is her body, floating with outstretched arms and legs through the sky, so that it remains unclear whether the myth will hold, and she will survive, or whether she has chosen a valiant mode of death – as a gesture of ethical self-sacrifice. For the viewers the closure of *Crouching Tiger Hidden Dragon* is both a beautification of death – and thus a protective image, covering over the real symbolic and bodily destruction of a person – as well as an eternal image – pointing towards death and averting death at one and the same time. Jen Yu's body is arrested in death, but because Ang Lee leaves her suspended between heaven and earth, and thus between life and death, between the ephemeral and the eternal, his cinematic representation also arrests death. Though death is explicitly invoked by virtue of her sacrificial jump as well as the staging of her floating body, it is also completely absent from the image.

From this, a further aspect of the contradictions underlying representations of death can be deduced. Though death is nowhere, it is, of course, at the same time everywhere, because death begins with birth and remains present on all levels of daily existence, each moment of mortal existence – after the cutting of the umbilical cord – insisting that its measure is the finality towards which it is directed. Death is the one privileged moment of the absolutely real, of true, non-figurative materiality as it appears in the shape of the changeability and vulnerability of the material body. On the one hand, then, it demarcates figurative language by forcing us to recognise that, even though language, when faced with death, is never referentially reliable, it also cannot avoid referentiality. Non-negotiable and non-alterable, death is the limit of language, disrupting our system of language as well as our image repertoire, even as it is its inevitable ground and vanishing point. On the other hand, signifying nothing, it silently points to the indetermination of meaning, so that one can speak of death only by speaking other. The impasse at issue can be formulated in the following manner: as the point where all language fails, it is also the source of all allegorical speaking. But precisely because death is excessively tropic, it also points to a reality beyond, evoking the referent that representational texts may point to but not touch. Death, then, is both most referential and most self-referential, a reality for the experiencing subject but non-verifiable for the speculating and spectating survivor.

Elisabeth Bronfen

Yet the numerous literary depictions of deathbed scenes also illustrate that representations of death not only attest to the fallibility of aesthetic language and the impermanence of human existence, but also confirm social stability in the face of mortality precisely by virtue of a language of death. The force of these narratives resides in the fact that in their last moments the dying have a vision of afterlife, while at the same time the aesthetic rendition of the deathbed ritual includes the farewell greetings from kin and friends and the redistribution of social roles and property that serve to negotiate kinship succession. Thus a sense of human continuity, so fundamentally questioned in the face of death, is also assured in relation to both ancestors and survivors. Indeed, as Walter Benjamin argues, death is the sanction for any advice a storyteller might seek to transmit. Speaking in the shadow of one's own demise, as well as against this finality, is precisely what endows these stories with supreme authority. The death of the Afro-American woman Annie (Juanita Moore) in Douglas Sirk's last Hollywood melodrama, *Imitation of Life* (1959), serves as a poignant illustration for the power of the deathbed scene. Having returned from Hollywood, where she has found her daughter, passing as a white showgirl, and having been unable to convince Sarah Jane (Susan Kohner) that she should come home with her, Annie finds herself fatally ill. Lying in her bed in her room in the home of Lora Meredith (Lana Turner), a Broadway star for whom she has been working for the past fifteen years, she comes to enact the solidarity between her surrogate family and the members of the black community, to which she has also always belonged. For the sentimentality unleashed by the anticipation of her death allows the survivors to reconfirm their alliances amongst each other, while Annie uses her final leave-taking not only to determine how her possessions are to be redistributed, but also to confirm the image she wants herself to be remembered by.

Indeed, the melodramatic power of this deathbed scene on the one hand feeds off the belief that in dying Annie has advice to give, which seems to be irrevocably inscribed by authority. On the other hand, Sirk uses this narrative convention so as to disclose the way in which the authenticity of emotions deployed here are nothing other than an imitation of life. Her doctor, her priest and the Afro-American butler Kenneth have all gathered around Annie's deathbed, while Lora has sat down by her. Between her head and Annie's we see a photograph of Sarah Jane smiling radiantly, propped up against the lamp on the bedside table. While Annie explains to those assembled how to pay her their last respects, how she wishes to divide up her possessions, she individually calls upon each one of her grief-stricken friends, and entrusts each with a particular concern. Although the most important task is entrusted to the entire group gathered around her bed, she has already turned her gaze from them, before she begins to explain to them how she imagines her funeral to be. If, in a prior scene, she had told Lora that our wedding day and the day we die are the great events of our lives, she now publicly elaborates on the dream she has been harbouring as her seminal fantasy all these years, namely that of finally reaching a more noble home than the one she has been inhabiting on earth. As Mahalia Jackson will sing in the scene of the funeral, immediately following upon this deathbed scene, the dying woman firmly believes that 'I'm going home to live with God.' The funeral ceremony she has planned down to the last detail, and for which she has saved all her life, is the ritual meant to show those surviving her one last time the fantasy which has allowed her to bear the unhappiness of her real living conditions as an Afro-American woman, relegated to the backrooms of the glamorous home of the Broadway star. From her deathbed she can now orchestrate this funeral, as though it were the one moment of power available to her. Lora responds to her description of her funeral with indignation and despair, and tries to convince her not to leave her. But, explaining 'I'm just tired Miss Lora, awfully tired', Annie leans back one last time onto her pillow with exhaustion and quietly closes her eyes.

According to Sirk, the 'no' with which Lora responds to her friend's sudden demise, is the one good line Lana Turner has in the entire film, the only moment in which her performance appears real. Indeed, Douglas Sirk has Lora call out in vain twice to the deceased, only to let her fall forward onto the bed in distress, thus lying next to the dead Annie. All we still see on the screen at the end of this scene is the face of Sarah Jane on the photograph, now framed by two mother figures, who have both turned away from her. With this mise-en-scène, Sirk undermines the grand emotions which Annie's speech aroused in the spectators, because the daughter, on whom her entire emotional life had depended, is not merely absent. Rather, her radiant smile is nothing but an image. Annie's conviction that her funeral will represent her proud transition into God's glory in worldly terms is as much a protective fiction as the boundless love for her daughter which she proclaimed on her dying bed. Furthermore, the emotional melodrama of this performance clearly exceeds its mark; its sentimentality is over the top in a programmatic way. For Sirk deploys an excessive mise-en-scène so as to illustrate that death exceeds any representation, even as it proves to be the catalyst for a grand performance, such as the funeral ceremony and the final restitution of the family, with which the *Imitation of Life* ends. Sarah Jane will return just in time for the funeral, will finally publicly acknowledge the Afro-American mother she has been trying all her life to deny, only to discover that the Afro-American community burying Annie excludes her. She will thus end up sitting in the car with a surrogate family – Lora Meredith, Lora's lover and her daughter – cut off from the dead mother she refused to acknowledge during her lifetime, yet also haunted by her spirit.

Representations of death, one can thus say, ground the way a culture stabilises and fashions itself as an invincible and omnipotent, eternal, intact symbolic order, but they can do so only by incessantly addressing the opposition between death and life. As the sociologist Jean Baudrillard argues, the phenomenon of survival must be seen in connection with and contingent upon a prohibition of death and the establishment of social surveillance of this prohibition. Power is first and foremost grounded in legislating death, by manipulating and controlling the exchange between life and death; indeed by severing the one from the other and by imposing a taboo on the dead. Power is thus installed precisely by drawing this first boundary, and all supplementary aspects of division – between soul and body, masculinity and femininity, good and bad – feed off this initial and initiating separation that partitions life from death. Any aesthetic rendition of death can be seen in light of such ambivalent boundary drawing. This is, of course, precisely Douglas Sirk's point at the end of *Imitation of Life*. After Annie's death, social power – notably the law of racial segregation dominant in the USA in the 1950s – is reinstalled. Her corpse and the boundaries drawn around it and in relation to it serve to renegotiate other boundaries, primarily involving the question who can be included and who must be excluded from both the cultural fantasy of the happy family as well as that of an allegedly intact black community.

Between the tropic and the real

Referring to the basic fact of moral existence, these representations fascinate because they allow us indirectly to confront our own death, even though on the manifest level they appear to revolve around the death of the other. Death is on the other side of the boundary. We experience death by proxy, for it occurs in someone else's body and at another site, as a narrative or visual image. The ambivalent reassurance these representations seem to offer is that, although they insist on the need to acknowledge the ubiquitous presence of death in life, our belief in our own immortality is nevertheless also confirmed. We are the survivors of the tale, entertained and educated by virtue of the death inflicted on others. Yet, although representations of death may allow us to feel assured because the disturbance played through in the narrative ultimately finds closure, the reader or spectator is nevertheless also drawn into the liminal realm between life and death, so that partaking of the fantasy scenario often means hesitating between an assurance of a reclaimed mastery over and submission before the irrevocable law of death. It is an ambivalent power that is attributed to the survivor – and spectator – of the death of another. But therein also resides the power of myth, which according to Barthes (1956) entails depleting a body of its historical context and raising it to the level of a mythic signifier. One might take the ending of Elvis Presley's first Hollywood film, *Love Me Tender* (1956), as an example of the power contained in such a figurative relationship between death and iconic resurrection. The film, set in the backwash of the Civil War, involves a fatal rivalry between the two Reno brothers. Both are in love with Cathy (Debra Paget), and when the older brother Vance doesn't return after the war, Clint (Elvis Presley) decides to marry her instead. Once his brother returns, of course, the fraternal feud requires the sacrifice of one of them for the family peace to be restored, and it is the younger one who literally takes the bullet. He dies in the arms of the woman, who had never really loved him, assured by the brother he implicitly betrayed that everything will be all right. The director Robert D. Webb then cuts to the funeral ceremony, during which we hear Elvis Presley's voice-over, singing the title song. So as to emphasise the fact that, though dead, the young cowboy lives on as an image in the minds of the survivors, the final shot of the film shows the family ranch, with an image of Elvis Presley, singing while strumming his guitar, superimposed on this emblem of family unity.

But of course the film image is more complex. As the story of this film's production history has it, both Elvis Presley's mother and his fans were horrified at the thought that their idol would die at the end of the film. Thus the producers came up with the idea of a 'singing corpse' superimposed over the closing images. The reel character Clint Reno dies, but not so that the real actor, Elvis Presley, can live on. Rather the power of the icon 'Elvis Presley' that was being installed with this first Hollywood film resided precisely in the fact that the actual historical person Elvis Presley came to die 'figuratively', in order to be re-born as a mythic creature – a dead body resurrected as a cinema image, greater than life, beyond the boundary that delimits normal mortals from celestial creatures; forever singing somewhere in a site above our heads. This was to be a fatal exchange, as we know, for the young hillbilly from Memphis, Tennessee, who was to suffer all his life under the 'image' that had made him an international star and a millionaire in 1956, but also so powerfully had frozen him into an icon.

One might surmise that any representation of death, therefore, also involves the disturbing return of the repressed knowledge of death, the excess beyond the text, which the latter aims to stabilise by having signs and images represent it. As these representations oscillate between the excessively tropic and a non-figurative materiality, their real referent always eludes the effort of recovery that representations seek to afford. It disrupts the system at its very centre. Thus, many narratives involving death work with a tripartite structure. Death causes a disorder to the stability of a given fictional world and engenders moments of ambivalence, disruption or vulnerability. This phase of liminality is followed by narrative closure, where the threat that the event of death poses is again reclaimed by a renewed return to stability. Yet the regained order encompasses a shift because it will never again be entirely devoid of traces of difference. Ultimately these narratives broadcast the message that recuperation from death is imperfect, the regained stability is not safe and the urge for order is inhabited by a fascination with disruption and split. The certainty of survival emerges over and out of the certainty of dissolution.

Ultimately, the seminal ambivalence that underlies all representations of death thus resides in the fact that, while they are morally educating and emo-tionally elevating, they also touch on the knowledge of our mortality, which for most is so disconcerting that we would prefer to disavow it. They fascinate with dangerous knowledge. In the aesthetic enactment, however, we have a situation that is impossible in life, namely, that we share death vicariously and return to the living. Even as we are forced to acknowledge the ubiquitous presence of death in life, our belief in our own immortality is confirmed. The aesthetic representation of death lets us repress our knowledge of the reality of death precisely because here death occurs in someone else's body and as an image or a narrative. Representations of death, one could say, articulate an anxiety about and a desire for death, functioning like a symptom, which psychoanalysis defines as a repression that, because it fails, gives to the subject, in the guise of a ciphered message, the truth about his or her desire that he or she could not otherwise confront. In a gesture of compromise, concealing what they also disclose, these fundamentally duplicitous representations try to maintain a balance of sorts. They point obliquely to that which threatens to disturb the order but articulate this disturbing knowledge of mortality in a displaced, recoded and translated manner, and by virtue of the substitution render the dangerous knowledge as something beautiful, fascinating and ultimately reassuring. Visualising even as they conceal what is too dangerous to articulate openly but too fascinating to repress successfully, they place death away from the self at the same time that they ineluctably return the desire for and the knowledge of finiteness and dissolution, upon and against which all individual and cultural systems of coherence and continuation rest.

Epilogue

As Stephen Greenblatt compellingly claims in his introduction to *Shakespearean Negotiations*, he was initially driven by the desire to speak with the dead, only to discover that if – in uncovering social energies that have culturally survived – he had wanted to hear one voice, he found himself confronted with many voices of the dead instead, and that if he has wanted to hear the voice of the other, he had heard his own voice resonating in this exchange of power as well, because the speech of the dead is not private property. To illustrate how we are haunted by culture in the sense of being haunted by the voices of the dead, or, put another way, in the sense that as cultural analysts we inevitably enter into an exchange with the dead – the dead text, the world it emerges from and which it survives, but also, of course, the different shapes its cultural circulation has taken on – I will conclude with a final poignant example of the haunting power of a conversation with the dead – Baz Luhrman's radically post-modern performance of Shakespeare's *Romeo and Juliet* (1996). In the final death tableau, we find precisely the oscillation between tropic and real that has been at stake in my discussion, even while Luhrman's performance of this haunting is complex. Not only are the two star-crossed lovers from the start haunted by a desire for death that is stronger than any desire for survival, with their erotic self-expenditure always already fatally marked, in part because it is a transgressive, forbidden love, but in part because it is so clearly suicidal.

Rather, any performance today of this text is itself haunted by the voices of the dead, which is to say the many performances preceding it that have turned this final death tableau into a cultural cliché. Faced with this problem, Baz Luhrman's solution is to address precisely what is at stake if one enters into the powerful exchange between life and death.

After Juliet has shot herself in the mouth, only to collapse on top of the corpse of Romeo, who had taken poison after finding his beloved lavishly laid out in state amongst candles and flowers in the church he had meant to marry her in, the two dead bodies are staged as an allegory for a beautification of, but also as a mixture of voices of, the dead. For Baz Luhrman replaces the Shakespeare text, spoken by Claire Danes and Leonardo diCaprio, which has now fallen silent, with an intonement of Wagner's 'Liebestod' motif that takes over the spoken text on the sound-track. At the same time Luhrman also conjoins the Shakespearian text with another convention of how unhappy endings might be depicted, namely the composite of all romantic scenes leading to this tragic resolution. Over the dead bodies of his star-crossed lovers Luhrman shows us the scenic moments that made up their romance, only to counter these highly tropic – one might even say kitsch – images with the images of their corpses as seen on TV. Indeed, one might read this shift from aesthetically staged corpses, to flashback images commemorating the scenes of their ill-fated love, to quasi-realistic depictions of the bodies the ambulance is taking away, as a deft transformation of real bodies (dead) into mythic signifiers (the Shakespearian figures that were always already fictional creatures), so as to signal that they can now live for ever as icons over and against their real death. This is a moment where the power of death as a moment of transfiguration is used to ward off the power of death's inescapable reality. The eternity of a fiction is pitted against real death, even while this is not just another idiosyncratic interpretation of Shakespeare's actual end, where Montague declares: 'I will raise her [Juliet's] statue in pure gold that whilst Verona by that name is known, there shall no figure at such rate be set, as that of true and faithful Juliet.' Rather, Luhrman performs his version of the monument, and does so significantly in reference to his own film; i.e. to the medium he has chosen to enter into a dialogue with death, with the dead Shakespeare's text and with the dead lovers of the story Shakespeare immortalised. The power in his cinematic images, as he negotiates the boundary between life and death, is that of commemorating and keeping eternal our memory of Juliet and Romeo, as tropes that both supersede and at the same time feed off a death they can gesture towards but never fully touch. Death, once more, proves to be both solitary and public; a supremely unique event yet also fully codified and always already culturally negotiated.

REFERENCES

Ariès, Philippe (1981). *The Hour of our Death*. Trans. by H. Weaver. Oxford: Oxford University Press.

Barthes, Roland (1981). *Camera Lucida*. Trans. by R. Howard. New York: Hill and Wang.

(1956). 'Myth Today', in Susan Sontag, ed., *A Barthes Reader*. London: Vintage, 1993, 93–149.

Bataille, Georges (1957). *L'Erotisme*. Paris: Editions de Minuit.

Baudrillard, Jean (1976). *L'Echange symbolique et la mort*. Paris: Gallimard.

Benjamin, Walter (1977). 'Der Erzähler', in R. Tiedemann and H. Schweppenhäuser, eds., *Gesammalte Schriften 2.2*. Frankfurt am Main: Suhrkamp, 438–65.

Blanchot, Maurice (1981). *The Gaze of Orpheus and Other Literary Essays*. Ed. P. Adams Sitney, trans. by L. Davis. Barrytown, N.Y.: Station Hill.

Bloch, Maurice, and Jonathan Parry, eds. (1982). *Death and the Regeneration of Life*. Cambridge: Cambridge University Press.

Bronfen, Elisabeth (1992). *Over her Dead Body. Death, Femininity and the Aesthetic*. Manchester: Manchester University Press.

Foucault, Michel (1977). *Language, Counter-memory, Practice: Selected Essays and Interviews*. Trans. and ed., D. F. Boucard and S. Simon. Ithaca: Cornell University Press.

Freud, Sigmund (1920). 'Beyond the Pleasure Principle' in J. Strachey, ed., *Standard Edition of the Complete Psychological Works of Sigmund Freud*, XVIII. London: Hogarth Press, 1955.

Greenblatt, Stephen (1988). *Shakespearean Negotiations. The Circulation of Social Energy in Renaissance England*. Berkeley: University of California Press.

Heidegger, Martin (1922). *Sein und Zeit*, 15th edn. Tübingen: Max Niemeyer, 1979.

Vernant, J.-P. (1991). *Mortals and Immortals: Collected Essays*. Ed. F. Zeitlin. Princeton: Princeton University Press.

6 The power of music

DEREK B. SCOTT

Perhaps the most popular perception of music's power is as a force acting upon or representing emotions such as love, hate, fear, joy and sadness. Because of its emotional impact, music also possesses a political power that can be exerted in the forging of national and social class identities. The British National Anthem offers a telling case study of such power. A subject frequently linked to politics is economics, and here again music exerts its might. The value of the music sector to the UK economy was an estimated £2.5 billion in 1995, according to a National Music Council Report (Eliff, Feist and Laing 1996: 5). Four record companies (BMG, Universal, Sony and Time-Warner) currently control over 80 per cent of the global record market. When the Spice Girls split up, the price of EMI shares fell immediately by 10 pence. Their recovery was helped by pop star Britney Spears. Alas, on 28 January 2002 EMI's shares fell again, this time slipping 14 pence after Deutsche Bank warned that the shareholders' dividend was likely to be halved. This was a consequence of the company's decision to pay one of its artists not to sing. The sales of Mariah Carey's latest album had been disappointing, but EMI were locked into a deal to fund several more of her albums, so offered her instead a pay-off of £19.6 million. Major record companies are dealing with stars whose financial value is greater than the gross domestic product of some small countries.

For an example of the sums of money that can be generated in the world of popular music, we can take the Irish rock band U2. They have sales in over a hundred different markets, and are involved in a maze of company deals negotiated by their manager Paul McGuinness. The *Sunday Times* revealed (3 June 2001, Section 9, p. 6) that the gross take from U2 sales for the year 2001 was estimated to be in the region of £210–230 million. Of that, some £135 million

Power, edited by Alan Blackwell and David MacKay. Published by Cambridge University Press.
© Darwin College 2005.

can be put down to record sales. Their American tour that year included 50 shows, each attended by an average of 20 000 people at around £60 per ticket, thus totalling £60 million. Their European tour – 30 shows with an average price of £31.50 and estimated ticket sales of 600 000 – accounts for £19 million. Sales of U2 authorised mechandise adds up to about £5.5 million in the USA and £2 million in Europe. On top of this approximately £7 million needs to be added, being the royalty payments from radio stations and from artists covering their songs. Of course, it would be wrong to forget the costs the band incurs: for example, they travel in a chartered Boeing 707 with two dozen staff, and another hundred travel by bus and truck.

Yet, it should not be forgotten among such material considerations that music's economic might has been harnessed to charitable causes such as the 'Feed the World' campaign. Nor should we forget that many people value music above all for its intellectual, spiritual or moral power. At the same time, however, there has never been a consensus as to whether music's spiritual power is greater than its morally corrupting power, or vice versa. This chapter explores the power of music as it is exerted in each of these different domains and, finally, examines a remarkable new insight into the power of music on the brain that has come to light in recent years.

It may be worth pausing at the outset to consider first how an audio engineer might measure music's power. When a friend of mine heard that I was to give a lecture on this subject, he immediately assumed that I would be talking about rates of energy flow, measured in watts. For instance, if we are looking for a scientific formulation, we should be aware that a 75-piece orchestra playing very loudly produces around 70 watts; in other words, barely enough power to light an average domestic room. Of course, having said 'very loudly', my friend would remind me that loudness is subjective, and that I should speak, rather, in terms of acoustic magnitude measured as sound pressure level (spl) plotted in decibels (dB). Then, I would need to ensure that I did not introduce into this chapter a confusion between sound power and sound pressure, the first of these being the square of the other. It is at this point that I have to confess that I am intellectually better equipped to discuss the power of music as conceived in terms of its effect upon our emotions and behaviour rather than in terms of watts. Fortunately, examples of music's power to represent human emotion are legion. A quick list will give some idea of the range: there is the youthful joy of the first movement of Mendelssohn's 'Italian' Symphony, the grief of the final chorus of Bach's 'St Matthew Passion', the terror of the *non confundar* section of Berlioz's 'Te Deum', and the passionate love music in Wagner's *Tristan und Isolde*. In Chapter 19 of *The Descent of Man* (1871), Charles Darwin remarked that, a single note of music could contain a greater intensity of feeling than pages of writing. However, he also noted perceptively that, though music affects every emotion, it 'does not by itself excite in us the more terrible emotions of horror, rage, etc.'.

These words on music and the emotions need to be balanced by considering the intellectual power of music. Some composers have at times wished to concentrate on solving musical problems rather than representing emotions. Here, we might consider such ingenious feats as the two string quartets (Nos. 14 and 15) by the French composer Milhaud that can be played either separately or simultaneously. Another example of this intellectual approach to music is the minuet from Haydn's Symphony No. 47, which the composer asks the performers to play both forwards and backwards. The intellectual dimension of music-making reached an extreme in the works heard at summer schools in Darmstadt in the 1950s. Some of the composers who gathered there were interested in the possibility of organising music systematically, so that all its elements, melodic, rhythmic, and even its dynamic changes between soft and loud could be regulated logically. Whether, in fact, music can ever achieve the rigour of philosophical or mathematical logic is, however, highly debatable.

Since I have no time in one chapter to cover every aspect of this subject, the rest of my essay will focus on the political and moral power of music, but I will have some words to say in conclusion about music and the mind from a more scientific perspective.

The political power of music

There are many aspects to the political power of music, and they involve questions of nationalism, class and identity. Music can be used to mark out territory, to control, threaten and oppress; and we need go no further than Northern Ireland to find evidence of that. A patriotic song may express patriotism to one community and triumphalism to another. In contrast, there are occasions when music seems to unite a nation in grief as happened with Elton John's 'Candle in the Wind' after the death of Princess Diana. In the United Kingdom, 'Nimrod' from Elgar's *Enigma Variations* has often performed this role, as in the United States has Samuel Barber's Adagio. An investigation into the changing uses and meanings of the British National Anthem over the past three centuries reveals the various ways in which the power of music can be harnessed to politics. And it seems especially appropriate to choose it as a case study, writing this in 2002 – the Golden Jubilee year of our present Queen.

'God Save the King' began as a musical oath of allegiance within a divided nation, but later became a symbol of British imperial power. The most significant date in the early history of what was to become the British National Anthem is Saturday 28 September 1745, when it was first performed before a London theatre audience. Exactly one week earlier, the 'Scotch rebels' fighting for Bonnie Prince Charlie and the House of Stuart had scored a major victory over the English fighting for the Hanoverian cause at the battle of Prestonpans. In response to this, Thomas Arne, the Musical Director of the Theatre Royal, Drury Lane, made an arrangement of an existing song calling for divine aid.

Initially, the song was a mixture of hymn and dance – its rhythm being that of a galliard. In the *Thesaurus Musicus* publication of 1744, which reproduces the earliest known engraved print of the tune, the dance element is still strong, the harmonic treatment lending a lightness of touch to the typical galliard rhythm of the tune. When we turn to Thomas Arne's arrangement of 1745, we find musical changes as well as an attempt to give the words greater specificity – 'God bless our noble King, God save Great George our King', instead of 'God save our Lord the King, Long live our noble King'. Arne alters the structure of the song by repeating each half of the tune (no repeats are marked in the 1744 version), introducing a solo and chorus antiphony that has sacred connotations (priest plus congregation). He adds decoration to the melody, but some of it, for example the grace notes added to the second line of the stanza, become firmly attached to the tune from now on. He replaces some of the plain harmony with what in his day would have been perceived as more colourful and emotional chords (the beginning of the second half of the tune), and he adds weight to the tune by increasing the number of chord changes. The harmonic treatment is now drawing closer to that of a hymn (with chord changes on each syllable) than a dance. Later versions of the song continue the move from the secular world of dance towards heartfelt devotion.

For a while, the song remained closely associated with the '45 Jacobite Rebellion; indeed, before the Jacobites had been defeated, a fourth verse was added referring to Marshal Wade's departure for Scotland on 6 October 1745.

O grant that Marshal Wade
May, by Thy mighty aid,
Victory bring.
May he sedition crush,
And like a torrent rush,
Rebellious Scots to crush:
God save the King.

It has often been assumed, and still is, that lines in the second verse (especially, 'Confound their politics, Frustrate their knavish tricks') were also written specifically with 'rebellious Scots' in mind, but evidence shows that they predate the events of 1745 (it is estimated in Dart (1956: 207), that the plate used in *Thesaurus Musicus* was engraved between 1735 and 1740). Percy Scholes (1954: 57) reproduced a photograph of an inscribed drinking glass that points to the possibility of the song's existing in some form in the late seventeenth century (see Figure 6.1). Ironically, this means it could have started out as a song for the Roman Catholic James II of the Stuart line. The indisputably anti-Scottish fourth verse given above did not survive long. It is too simplistic, of course, to view the '45 Rebellion as a matter of Scots *versus* English; there were many Stuart sympathisers in England at that time. Hence, the rapid spread of London theatre performances of 'God Save the King' served a clear function, that of teasing out the 'disloyal' and strengthening the commitment of the rest.

As years passed, it began to be realised that this was not just one of many patriotic songs tied to particular political events and destined to become a historical curiosity, but a new breed of song – the world's first National Anthem, in fact. However, the term 'national' would not, in this period, have been interpreted to mean that a particular ethnic character was discernible in its musical style; it merely meant that its sentiments were resolutely patriotic. Other countries must have held similar views, given that the tune of 'God Save the King' served as a vehicle for patriotic songs in America, Denmark, Prussia, Italy, Russia, Sweden, Switzerland and elsewhere in the eighteenth and nineteenth centuries. Liechtenstein still uses it for its National Anthem 'Oben am jungen Rhein', and that despite a declaration by the Chancellor of the Exchequer in the British Parliament in 1931 that 'only the tune itself' and not the words constituted the British National Anthem.

In the nineteenth century, 'God Save the Queen' became a symbol of British imperial unity. It is especially evident in the celebrations that took place during

FIGURE 6.1 Jacobite drinking glasses.

Queen Victoria's Golden Jubilee of 1887. At one concert that year, which was also the thirtieth anniversary of the Indian Mutiny, it was sung in five different Indian languages. The National Anthem was everywhere during Golden Jubilee year, resounding whenever matters regal or imperial needed to be evoked. In fact, someone even made a bustle for Queen Victoria that played 'God Save the Queen' whenever she sat down – ensuring, presumably, that everyone remained standing even when the Queen was seated.

Because the National Anthem is a symbol, it is difficult to change any of its constituent parts without a consequent effect on its meaning. For example, the effect of changing its harmonies might be compared to changing the blue in the Union Jack to purple. The *Monthly Musical Record* warned against re-harmonisations in 1937, suggesting that they made people feel self-conscious, which is 'the enemy of common purpose'. The rhythm is crucial to the tune's recognition. Almost everyone in the UK can spot 'God Save the Queen' from the tapping of its rhythm alone without being given any melodic clues to its identity. On the other hand, I have made no changes to the actual notes of the tune in the version below, yet the rhythmic alterations result in its being almost impossible to recognise.

Nothing above should be taken as implying that nobody has ever tried to alter the National Anthem, either to improve on it or to subvert it. Hundreds of new verses written to its tune exist. There were several competitions last century to improve on the words – for example, those organised by the Royal Colonial Institute (1918), the *Morning Post* (1924), the Poetry Society (1935) and the *Spectator* (1952). Parodies can be found as early as 1748, when there was one written in praise of Westminster Fish Market. A parody dating from the time of the French Revolution runs:

> To the just Guillotine
> Who shaves off heads so clean
> I tune my string!
> Thy power is so great,
> That ev'ry Tool of State,
> Dreadeth thy mighty weight,
> Wonderful Thing!

Shelley wrote his 'New National Anthem' in praise of the Queen of Liberty to go to the same tune.

At the height of the working-class Chartist Movement in 1848, the meaning of 'God Save the Queen' began to shift again as it took on more of the character of a *class* anthem than a *national* anthem. On the day of a massive Chartist meeting on Kennington Common, a Philharmonic Concert was scheduled. It was reported that during the performance of 'God Save the Queen' the audience (who would have been, in the main, wealthy middle-class subscribers to the concert series) began cheering and waving their hats and handkerchiefs at the line 'Confound their politics'. This is probably the first occasion on which the National Anthem was directed against those – social reformers, trades union sympathisers, etc. – whom the British Prime Minister Margaret Thatcher memorably described in the 1980s as 'the enemy within'.

In 1891 it was reported that some people had been heard to hiss at the National Anthem during the National Eisteddfod in Swansea, Wales. Fifty years on, the musicologist Percy Scholes witnessed two Welshmen in Aberystwyth being arrested when a scuffle ensued after they had refused to take their hats off or stand for the National Anthem. These incidents show that for some Welsh people 'God Save the Queen' has for long been perceived to be an English rather than British National Anthem. Some sporting events now make use of alternative anthems. For example, at the Rugby Union Six Nations Championships, Wales is represented by 'Hen wlad fy Nhadau' ('Land of My Fathers', 1856) and Scotland by 'Flower of Scotland', a twentieth-century folk-style ballad that has completely overtaken Scotland's official anthem 'Scots Wha Hae' (1793) in popularity.

Sometimes, people have been stopped from performing the National Anthem, as happened in Ireland in 1905, when students at Dublin University prevented the organist from playing 'God Save the King' at a degree ceremony. On other occasions, the problem could be the opposite, that of trying to win over a reluctant performer: in Canada in 1908, a man in Toronto murdered his wife because she would not sing 'God Save the King' on Empire Day (the incidents referred to may be found in Scholes (1954: 213 and 221–4)).

England for many years carefully coached its young people in the performance of the National Anthem: for instance, schools were teaching their pupils to sing it for Queen Victoria's Diamond Jubilee of 1897; and in March 1916 the London County Council ordered an inquiry into which version might be most suitable for school use. By 1960, however, the Minister for Education was merely stating, 'I certainly hope that all children do learn the National Anthem' (Eyck (1995: 21)). Then, in the following decade, it became evident that even in England, and during the most carefully organised of royal celebrations, things could go awry. For the Queen's Silver Jubilee of 1977, the punk rock band the Sex Pistols released an anarchic record entitled 'God Save the Queen', and

Derek B. Scott

enjoyed considerable success in the popular music charts despite the BBC's instructions that the song should be given no air time on radio or television. Moving closer to the present, there was a sharp reminder in June 1997 that the National Anthem can still play a well-worn role as an accompaniment to aggressive and chauvinist behaviour on the part of the Queen's loyal subjects. In that month, three young British marines were fined in a Cyprus court for stripping off and singing 'God Save the Queen' in a public square.

England does, of course, possess other well-known patriotic songs. Political parties during election campaigns shy away from using 'God Save the Queen', but have fewer qualms about turning to 'Land of Hope and Glory' (Benson/Elgar, 1902) or 'Jerusalem' (Blake/Parry, 1916). 'Rule, Britannia' (Mallet/Thomson/Arne, 1740) is, sadly, too implausible to be used for political rallies these days, and is mainly consigned to historical pageants or the high jinks of the 'Last Night of the Proms' at the Royal Albert Hall. It would also be best not to dwell upon the cruel fate in 2001 of that important symbol of Englishness celebrated in a patriotic song almost contemporary with 'Rule, Britannia' – 'The Roast Beef of Old England' (Fielding/Leveridge, 1734). Regrettably, fortune has failed to smile on England where the sanity of its cattle is concerned. The last attempt to create another National Anthem really belongs to 'There'll Always Be an England' of 1939 (Parker/Charles). There is, however, a backward-looking quality to its words which means the emotion it generates tends to be primarily nostalgia for a lost England of thatched cottages and country lanes. Moreover, the obviousness of its militaristic march rhythm and fanfares cannot equal that peculiar mix of spiritual supplication and worldliness possessed by 'God Save the Queen' and heard, for example, in the moving performance recorded by Dame Clara Butt during 1917, a time of carnage in Europe (the Passchendaele offensive of that year cost the lives of over 280 000 British soldiers).

The moral power of music

A survey of music's moral power would include its religious power, demonic power and erotic power. I shall make reference to these, but intend to concentrate on music's moral power as it was perceived in the nineteenth century, in the light of what are often referred to as 'Victorian values'. Holman Hunt's painting *The Awakening Conscience* illustrates the spiritually transforming effect of music (see Figure 6.2). The pianist has been idly playing Tom Moore's

FIGURE 6.2 Holman Hunt, *The Awakening Conscience.*

ballad 'Oft in the Stilly Night' to his mistress. Unfortunately for his libidinous intentions, instead of its acting as a stimulus to seduction, the music has awakened her moral conscience.

Victorian middle-class values included thrift, self-help and hard work, but, where music was concerned, the key value in asserting moral leadership was respectability. It allowed a moral stand to be taken against certain aspects of working-class behaviour, especially drunkenness and immorality. Respectability is not enforced from on high, however; it operates as part of a consensus won by persuasion. The fight for respectability was one that religious organisations were eager to support. Non-conformism was a major force behind English choral music in the nineteenth century. The Methodists, for example, had introduced congregational singing in the previous century, and a desire to encourage education and 'improvement' made them strongly committed to sacred choral music. London's Sacred Harmonic Society, founded in 1832, began as a non-conformist organisation.

The rational and the recreational were linked together in the sight-singing movement, even if the singing was not from conventional notation. Joseph Mainzer, the author of *Singing for the Million*, John Hullah and, last on the scene, John Curwen each offered competing methods to the singing classes, the latter promoting the Tonic Sol-Fa method devised by Sarah Glover, a teacher in Norwich. The publishing house Novello, set up in 1811, took over the publication of Mainzer's *Musical Times and Singing Circular* in 1844, by which time the firm specialised in producing cheap musical editions, especially of Handel, for choral societies. Enormous triennial Handel Festivals, involving up to 2000 performers, took place from 1857 in the Crystal Palace at Sydenham, South-East London.

The working class was thought to need 'rational amusement' such as choirs. It was not a cynical exercise in control: in their own lives the middle class were committed to self-improvement by going to concerts, buying sheet music and performing it at home. From the 1830s on, pianos were found in middle-class homes, and girls were expected to learn to play them. A belief in the moral power of music was all-pervasive: 'Let no one', admonished the great champion of the improving powers of music, the Reverend Haweis, 'say the moral effects of music are small or insignificant' (1871: 112). Arthur Sullivan was of the opinion that music could suggest no improper thought.

In the 1850s, the sale of refreshments was permitted on Sundays in certain London parks to coincide with military band performances. It met with strong opposition from those who wished to guard Sunday's importance as a religious day and who feared, also, that the excitement of listening to band music would trigger civil disturbance. On the other hand, the right kind of music (say, concert or sacred music), in the right surroundings, was thought to act as 'a civilising influence to which the lower classes were particularly responsive' (see Mackerness (1964: 185–6 and 202)). This accords with the conviction behind Matthew Arnold's *Culture and Anarchy*, published in 1869, that culture is needed to save society from anarchy. Culture for Arnold is not a broad term: he spares no time on the music hall.

In the 1880s in London, music halls were keen to promote their respectability and 'superior class' of audience (see Russell (1997: 93)). In the next decade in New York, 'vaudeville' became the accepted term for respectable variety entertainment. Tony Pastor, often described as the 'Father of Vaudeville' (although he preferred the term 'variety') began as a circus clown, but started appearing in variety shows on Broadway and in the Bowery in the 1860s. From 1881, he was staging 'high class' variety shows at his own theatre on 14th Street. The vaudeville shows to be seen at Proctor's theatre on 23rd Street were described by a New York judge as being of the kind 'that keeps a family together' (quoted in Marston and Fuller (1943: 48)).

It was meaningless, of course, if the entertainment was respectable but the venue not. Concern about prostitution in theatres and music halls grew in the second half of the century. Alcohol consumption was another threat to morals and respectability, and music was used as a medium of persuasion by temperance groups in London and New York that promoted songs portraying the destructive effects of drunkenness on the home and family. Some of these pulled no punches in getting their message across: what hard-bitten drinker could ever forget the image of a tiny orphan girl standing alone singing that temperance classic 'Father's a Drunkard and Mother Is Dead' (Stella/Parkhurst, 1868)?

> Out in the gloomy night, sadly I roam,
> I have no Mother dear, no pleasant home;
> Nobody cares for me – no one would cry
> Even if poor little Bessie should die.
> Barefoot and tired, I've wander'd all day,
> Asking for work – but I'm too small they say;
> On the damp ground I must now lay my head–
> *'Father's a Drunkard, and Mother is dead!'*

The efforts made in the interests of respectability were not always an unqualified success and, sometimes, failure appeared in an unexpected quarter. In 1899, the American *Musical Courier*, with reference to ragtime, proclaimed: 'A wave of vulgar, filthy and suggestive music has inundated the land' (quoted in Whitcomb (1986: 16)). The irony was that ragtime idiomatically suited the instrument most imbued with domestic respectability, the piano. However, the flipside of this was that pianos were also commonly to be found in New York's brothels and honky-tonks.

In the twentieth century, certain kinds of popular music continued to cause moral anxiety. The American Ragtime Octette's 'Hitchy Koo!' (Muir/ Abrahams/Gilbert) was the forerunner of many nonsense songs of the twentieth century suspected of being indecently suggestive: in 1956 the BBC banned Gene Vincent's 'Be-Bop-a-Lula' (Davis/Vincent) for that reason. It may be surprising that the same social values that permitted the smutty innuendoes of George Formby were less accommodating towards scat singing (a form of improvised singing on nonsense syllables), the reaction to which ranged from suspicion to outrage. To take an extreme case from Nazi Germany, the records of British trumpeter and bandleader Nat Gonella were banned there and scat singing was a criminal offence. In the 1990s in Afghanistan, the Taliban went even further and banned all music.

In 1920s Britain, ideas of propriety and the dance were inevitably intertwined with issues of respectability, class and status. Modern ballroom dancing may be dated back to an influential manual of that name issued by Vernon and Irene Castle in 1914. The Castles were pioneers of modern dance, and of the long-lived fox trot in particular. Although Vernon was English, it was in the USA that he and his wife found fame, where the celebrated band of the black American James Reese Europe played for their *thés dansants* (tea dances). New dance halls were being opened at the end of 1919 throughout the length and breadth of Britain, despite condemnation of the new dances by the clergy and other moral guardians who, by way of contrast, were happy to direct those with an appetite for dancing towards the English Folk Dance Revival.

In 1926 there was an attempt by some politically motivated moralists to reassert 'traditional values' in opposition to dance-band music, and to forge a link between musical propriety and the 'democratic spirit'. They spoke of a musical 'folk' heritage, once part of the fabric of national life, but now being neglected in favour of fashionable but inferior products. The year 1926 was turbulent: in Rome, Mussolini addressed 60 000 blackshirts; in Britain, there was a lengthy miners' strike (following a lock-out) and a nine-day General Strike. It was as the miners' strike was collapsing, after its seventh month, that the first *Daily Express* Community Singing Concert was held in the Royal Albert Hall. The proprietor of that newspaper, Lord Beaverbrook, had conceived the idea of a 'nation united in song' (an idea he got, perhaps, from hearing crowds singing at football matches).

The songs he published in the Daily Express Community Song Books are representative of British nationality as a whole; included, for example, are Welsh songs, like 'All Through the Night' ('Ar Hyd y Nos'), and Scottish songs, like 'Loch Lomond'. These songs must have taken on the solemnity of hymns as they were sung by massed voices accompanied by pipe organs in city halls up and down the land. In 1927, Feldman was publishing band arrangements with titles like 'Community Land'. These would offer a selection of 'community songs', determinedly avoiding anything resembling a fox-trot rhythm. More recent songs were tolerated if they had associations with patriotism, either directly, like 'Land of Hope and Glory', or indirectly, such as those sung by the troops during the First World War, for example 'It's a Long, Long Way to Tipperary' (Judge/Williams, 1912). Community Song Book No. 3 was entitled *Songs That Won the War.*

Few in the twentieth century would have ascribed to the typical Victorian view about the moral purity of music. In the Edwardian period, Henry Draper's painting *Ulysses and the Sirens* (1909) shows how even the most heroic of men can be tormented by bewitching voices (see Figure 6.3). Homer informs us that men who listened to their enchanting song were never more welcomed home by their wives and happy little children, but ended up instead joining a pile of rotting corpses on the Sirens' isle. In Homer's *Odyssey*, there are only two Sirens and they remain singing in their meadow. Draper has decided the story calls for a bit more drama as well as additional female flesh.

One of the biggest challenges to the view of music as pure and wholesome is its ability to communicate an erotic charge. This was something few Victorians wished to dwell upon, although Darwin was of the opinion that the vocal organs in animals had originally developed in relation to the propagation of the species. To discover how an erotic charge is accomplished in non-vocal music, we might consider the music of two stripteases, that which appears in Richard Strauss' opera *Salome* composed in 1905, and David Rose's big band hit of 1962 entitled

FIGURE 6.3 Draper, *Ulysses and the Sirens*.

'The Stripper'. The eroticism of Salome's 'Dance of the Seven Veils' is encoded in the sensual richness of a huge orchestra, the exotic, quasi-Oriental decoration of its melodies, and the devices of *crescendo* and quickening pace. Yet, equally important here, and despite its incongruity with the court of King Herod, is the Viennese waltz that lies just beneath the surface, providing connotations of *fin-de-siècle* decadence for an early twentieth-century audience.

David Rose's 'The Stripper' also needs to be placed in cultural and historical context. In the late 1920s, Duke Ellington's band had satisfied the expectations of the white patrons at New York's Cotton Club by developing a range of instrumental techniques for representing cries, howls and growls, and serving them up as supposed 'jungle music'. Erotic singers like Mae West then appropriated the smears, bent notes and growling plunger-mute effects, thus causing associations of the wild and the primitive to pass over to themselves. Next, having gained connotations of wild, predatory female sexuality, these effects (which are now heard as female cries, purrs, moans, groans and breathless gasps) could return as a highly charged instrumental eroticism. In 'The Stripper' we hear, for example, quasi-vocal slides on trombone and a wailing *tremolando* on a jazzy 'blue' note followed by 'jungle' drums.

To conclude this section on the moral power of music, I will add a few comments on the subject of music and the demonic with reference to the Danish philosopher Kierkegaard, who believed that music was not part of a moral or ethical domain. He wrote a long essay on the demonic in Mozart, focussing on what he regarded as the demonic absorption in desire of the character Don Giovanni in Mozart's opera of that name. Kierkegaard advances the idea that music, unlike language, is an imperfect medium, which causes it problems in attaining the immediately spiritual as its object. This is not to claim that music is the work of the devil, but that music is available and suited to the sensuous and demoniacal because, while being of a much greater degree of abstraction than language, it is able better to depict mood and passion. For Kierkegaard, Christianity is the *Word* made flesh, but music is a communicative medium that does not have the same ethical and religious relationship to Christianity – in short, music is not part of a moral or ethical domain. He refers to the many stories and legends in which mermaids or mermen put persons under musical spells. Breaking such spells often requires the music to be played backwards. Similarly, the demonic character inverts meanings and negates the truth. The demonic character is, for Kierkegaard, not the opposite of the good but, rather, one that despairingly defies the good in fear of what is ultimately the source of his or her personal salvation (thus resembling the condition psychologists call 'resistance').

The power of music on mind and body

Since time immemorial, music has been credited with extraordinary power over the body, and a power not just restricted to the human body if we are to give any weight to the legend of Orpheus taming the wild beasts (see Figure 6.4). Music has been helpful in many kinds of therapy: established and successful use of music therapy may be found in geriatric care, in alleviating mental health problems, and in encouraging and developing communication skills in autistic children. Here I want to touch on two recent developments.

The first is Medical Resonance Therapy Music, a term coined in the late 1980s for a new branch of medicine with implications for a variety of research areas, including dermatology, headaches, stress and intensive care medicine. The composer–creator of Medical Resonance Therapy Music is Peter Hübner,

FIGURE 6.4 Roman mosaic of Orpheus

who decided that the 'natural laws of harmony' could be harnessed to relieve psycho-physiological manifestations of stress. He prepared his 'medical preparations' on some dozen or so CDs. These are not to be listened to in the normal sense, being intended as a kind of medicine in sound rather than as entertainment. Currently, in Byelorussia, where scientists have been convinced of its value, all women with high-risk pregnancies are given Medical Resonance Therapy Music. The theory behind this type of therapy is that the body and its functions have a natural harmonic order and illness results if it is disrupted. Medical Resonance Therapy Music, it is claimed, helps restore order with a diet of 'scientifically based harmonic music medicine' (information from www.medicalresonancetherapymusic.com and www.digipharm.com). The idea of there being a music that is not meant to be listened to is not new. The origins of Muzak were in attempts to find a kind of music that would increase the productivity of workers by creating a feeling of well-being and vitality. Clearly, output would not increase if those workers started becoming distracted by listening to the music.

The second example I want to give of music's power to act on the body comes from neuroscientists researching into music and the brain. Gordon Shaw and Frances H. Rauscher working at the University of California, Irvine, in the 1990s, demonstrated that listening to Mozart's Sonata for Two Pianos, K448, for 10 minutes improved the spatial-temporal reasoning skills of a group of 84 students for approximately 1 hour (*Nature*, Vol. 365, 1993, p. 611). Such skills allow us to envisage patterns in space and time, as we need to do, for example, when playing chess. They are put to important use by engineers and architects in grappling with problems of proportion and geometry. A few years later, Shaw and Rauscher took 78 schoolchildren, divided them into three groups, and gave one group piano lessons, another singing lessons, and the remaining group training on computers. Comparing scores in tests taken before the lessons with tests completed six to eight months later, they found a 34 per cent improvement in the piano-trained children (*Neurological Research*, Vol. 19, 1997, pp. 2–8). At a similar time, Martin F. Gardiner published a study (*Nature*, May 1996) showing that schoolchildren who received a specialised musical training, as against a general musical training, were significantly ahead in their mathematical skills after a period of seven months.

The composer Mozart frequently emerges as someone whose music has a remarkable effect – so much so, that we have to beware of infringing copyright in describing it: there is now a registered trademark, The Mozart Effect®, owned by a resource centre in St Louis, USA. The website www.mozarteffect.com explains that it is 'an inclusive term signifying the transformational powers of music in health, education, and well-being'. Don Campbell, whose work is promoted by the centre, coined the term for the title of the first of his nine books dealing with the subject. Various recordings are available; they include a box of five cassettes entitled 'The Power of Music'. These, it is promised, will improve your IQ as well as your emotional intelligence (however that may be assessed). Among other wonderful benefits, you will learn how to use 'your morning commuting time to improve your entire business day'.

So, do we have conclusive evidence that a Mozart effect exists? If there is such an effect, does it rely on an arousal created by enjoyment of the music?

That seems to be contradicted by an experiment in which separated groups of rats were exposed when unborn and for 60 days afterwards to either white noise, minimalist music by Philip Glass, Mozart's K448 or silence. They were all then tested to see how well they negotiated a maze. The Mozart group won hands down – or perhaps I should say claws down – running through the maze faster and more accurately (*Neurological Research*, Vol. 20, 1998, pp. 427–32). Mozart's K448 has also proved remarkable in its effect on epilepsy. In one experiment it even produced an improvement in a patient who was in a coma, thus seeming to indicate that the effect is separable from 'music apprecia-tion' (*Clinical Electroencephalography*, Vol. 29, 1998, pp. 109–19, and Vol. 30, 1999, pp. 44–5). The next question has to be: is the effect solely tied to Mozart's music? Recent experiments indicate that it is not specific to Mozart's music, but insufficient comparative studies have been done to define what musical charac-teristics are necessary to produce this effect. Much energy has gone into trying to understand *how* K448 works on the brain rather than *why* this piece works.

And now, to take a quick glance at the other side of the coin, I should ask: what about those occasions when music seems to have the opposite of a healing power – for instance, when you're being driven mad, trying to stop humming a particular tune over and over again? The *Sunday Times* (23 Dec. 2001, Section 1, p. 20) reported that Professor James Kellaris of the University of Cincinnati had questioned 1000 people with experience of this problem. A com-puter game theme had remained in one person's head ever since first hearing it in 1986. Kellaris' theory is that certain combinations of simplicity, repetition and 'adrenaline-inducing jaggedness' create a cognitive itch that might be fan-cifully regarded as a mental mosquito bite. You scratch this musical mosquito bite by going over the tune again and again in your mind. At the risk of dam-aging the reader's peace of mind, I'll cite a variety of examples of these 'sticky' tunes – though I would advise against singing them. 'Yellow Submarine' by the Beatles appears to be one of the biggest culprits; and others are 'We Will Rock You' by Queen, 'Bad' by Michael Jackson, 'YMCA' by Village People, 'Follow the Yellow Brick Road' from *The Wizard of Oz*, and the 'Harry Lime Theme' from *The Third Man*. Even Mozart is guilty: some people find it impossible to get *Eine kleine Nachtmusik* out of their heads. Nevertheless, the Los Angeles clinical psychologist John Durrant thinks these 'sticky tunes' are a particular curse of our modern civilisation. He suggests a cold shower; but, for those who prefer a less dramatic solution, another remedy that has been put forward is cinnamon bark, which evidently acts as a sedative on the musical parts of the brain.

'Music has a wonderful power', proclaimed Darwin in his book *The Expression of the Emotions in Man and Animals* (1872). It is with some degree of irony that this survey of the power of music ends by offering advice on how to defend oneself against that power. We ought, perhaps, to conclude from this that all forms of power have positive and negative aspects, and the art of music is no exception.

FURTHER READING

Boyd, M. (2001). 'National Anthems', in S. Sadie, ed., *The New Grove Dictionary of Music and Musicians*, 2nd edn. London: Macmillan, Vol. XVII, 654–87.

Campbell D. (1997). *The Mozart Effect*. New York: Avon.

Jenkins, J. S. (2001). 'The Mozart Effect', *Journal of the Royal Society of Medicine*, Vol. 94 (April), 170–2.

Krummel, D. W. (1962). 'God Save the King', *Musical Times*, Vol. 103 (March), 159–60.

Scott, D. B. (2001). *The Singing Bourgeois: Songs of the Victorian Drawing Room and Parlour*, 2nd edn. Aldershot: Ashgate.

Viadero, D. (1998). 'Music on the Mind', *Education Week*, 8 Apr.

REFERENCES

Cliff, D., Feist, A., and D. Laing (1996). *The Value of Music: A National Music Council Report into the Value of the UK Music Industry*. London: National Music Council.

Dart, T. (1956). 'Maurice Greene and the National Anthem', *Music and Letters*, Vol. 37, No. 3, 205–10.

Eyck, F. G. (1995). *The Voice of Nations: European National Anthems and Their Authors*. Westport, Conn.: Greenwood Press.

Haweis, H. R. (1871). *Music and Morals*. London: Longmans, Green, 1912 reprint.

Mackerness, E. D. (1964). *A Social History of English Music*. London: Routledge and Kegan Paul.

Marston, W. M., and J. H. Fuller (1943). *F. F. Proctor Vaudeville Pioneer*. New York: Richard R. Smith.

Russell, D. (1997). *Popular Music in England, 1840–1914*, 2nd edn. Manchester: Manchester University Press.

Scholes, P. A. (1954). *God Save the Queen! The History and Romance of the World's First National Anthem*. London: Oxford University Press.

Whitcomb, I. (1986). *After the Ball: Pop Music from Rag to Rock*. New York: Limelight Editions, 1st published 1972.

7 Power in society

TONY BENN

Most people in the world feel powerless. There are three key political questions: 'What's going on?', 'Why?' and 'What can we do about it?' But if you ask 'What's going on?', people say, 'I'm not quite sure.' 'Why?' 'Well, nobody's told me.' 'What can we do about it?' 'Probably not very much.' That sense of powerlessness explains an awful lot of what we talk about as apathy, cynicism and so on, and this is what I want to address. I want to look at some of the sources of power, what they are, how they operate, who controls them, and ask the question, 'How can we get some influence over the powers that exist in order to improve our own lives?'

The most obvious example of power is power by conquest. Julius Caesar tried to get us into the European Union in 55 BC, and we still use the penny, which was a Roman coin. Margaret Thatcher was not the first Iron Lady – Boudicca killed 7000 Romans and raised the Men of Essex in order to deal with the Treaty of Rome. Power by conquest was the basis of all the great empires of the past. The British Empire is one example – when I was born in 1925, 20 per cent of the population of the world was governed from London. Now we live in an American empire. There are certain consequences of that, which impinge upon the way our politics are conducted.

We also live in a monarchy, and monarchy is hereditary. Although you might not think it matters very much, in order to serve in Parliament you have to give an oath: 'I swear by Almighty God that I will bear faithful and true allegiance to Her Majesty Queen Elizabeth II, her heirs and successors, according to law.'

This lecture was delivered in Cambridge on 18 January 2002, and makes several references to the political situation at that time. In order to retain the contemporary character of the lecture, the editors have not made changes to reflect subsequent events, including the 2003 invasion of Iraq.

Power, edited by Alan Blackwell and David MacKay. Published by Cambridge University Press.

Now I am republican; I think we should elect our head of state. So in order to serve in Parliament I've had to tell 17 lies under oath. You may not think it matters, but perhaps if we're the high court of Parliament we should adopt the other oath used by courts: 'I swear to tell the truth, the whole truth and nothing but the truth.' That might be an improvement in the House of Commons!

We still have a House of Lords. The current Prime Minister very boldly suggested that 20 per cent of them be elected. He will pick the list of course, so it would also be 100 per cent appointed. If you scratch the surface, class in Britain is based on the old idea: it's the landlords and serfs, and power and authority comes from above. The more I think about it, the more I think that the culture of British politics is that you owe your loyalty to the guy above you and not to the people who elected you. Loyalty has replaced, or is intended to replace, solidarity – what you feel for the people you represent – and that's very profound.

The second source of power is power by ownership, because wealth and power are indivisible. If you are rich, you get a lot of power; if you've got a lot of power, it's much easier to get a lot of money. Ownership as a route to political power was demonstrated very vividly when William the Conqueror arrived in 1066 and took over the land. That's what the Robin Hood story is all about. The feudal system was all about the ownership of land. When I was thinking about privatisation a year or two ago, I went to the House of Commons library and said, 'I want to introduce a bill to repeal the Enclosure Act', which took all the common land and handed it to the farmers. They said, 'What do you mean, repeal the Enclosure Acts? There are 10 000 of them.' Of course the land was handed over to private owners. Later we had the Industrial Revolution, the development of British capitalism, which Marx described, and now we have multinational companies which are more powerful than nation states. Having dealt with some of them myself, Esso, the oil companies and Ford, and so on, you were very well aware that they recognised that they had power. I remember when Henry Ford came to see me in my office – it was the grandson of the founder of course – I felt as if the emperor had visited a parish councillor and was telling me what we had to do. The gap between rich and poor in the world is very, very great – five hundred dollar-billionaires have the same income as half the population of the world put together. It is inconceivable that you could have long-term peace and stability with a division between rich and poor as wide as that.

The other thing about money by ownership, business money, is that it is used now to buy political power. The recent Mayor of New York, Bloomberg, spent $93 per vote for the votes he got. There must be a lot of people in New York who would have preferred to have the $93! That is on a massive scale. Enron, before their auditing scandal, funded both the Democrats and the Republicans, and apparently even put some money into the Labour Party. A former Governor of Ohio, an old friend of mine called Jack Gilligan, said, 'You'll never have democracy in America when big business buys both parties and expects a pay-off whichever one wins', and I think that was a very profound remark.

Another root of power is power by faith. The religions of the world have a great deal of power. Judaism, of course, not only retains a very powerful force for individuals today, but being the chosen people with a promised land gave them a claim, they believe, to territory. I was talking to a Palestinian who said he didn't know God was an estate agent – but that is the basis of the Jewish claim. Of course Christianity is a very powerful religion in many places, as is Islam. The thing about a faith is that it begins with a commitment to right against wrong, but faith quickly becomes structured and in no time at all the fire in your belly, which leads you to express your faith, becomes a fire you use to burn heretics, which is a very different sort of fire. In 1401, the House of Commons passed an Act, the Heresy Act, which said that any lay person (other than a priest) who read the Bible was guilty of heresy and should be burned at the stake. The Bible was seen by the Church as such a revolutionary document that if it got into the hands of the wrong people, who knew what would happen? And they were right to worry. In the old days the bishops would say to people, 'We know it's a very unfair world, but if only the rich are kind and the poor are patient, it'll be alright in heaven when we're dead.' And people said, 'Well that's marvellous news, bishop, could we have it while we're still alive?' That demand led to a great deal of social unrest.

The Christian church has used a lot of military force to retain such power. The Pope sent two armies to Britain to crush Pelagianism, the heresy founded by the British-born monk Pelagius, who said that justification is by works – it's what you do that decides your future – not justification by faith. That was a very dangerous idea as well. It was for this power that Henry VIII nationalised the Church of England. It is our oldest nationalised industry, and the Prime Minister still appoints the Archbishop of Canterbury. The King had the very clever idea that if you had a nationalised Church of England then there'd be a priest in every pulpit, in every parish, every Sunday, saying 'God wants you to do what the king wants you do', which was a very clever way of reinforcing your temporal power with a bit of spiritual support.

Another element of power is one that I can only call religious fundamentalism. I think the most powerful fundamentalism in the world today is the worship of money – much more powerful than Islam or Christianity or Judaism or Hinduism. Today the banks are bigger than the cathedrals and the temples. Even these religious fundamentalists have their own paramilitaries. The Mafia make money just by shooting somebody, which is the simplest way of making a profit. You don't even have to run a successful business, but it relies on the idea that the worship of money, the acquisition of money, is all that really matters. On television, we have business news every hour, but why can't we have other information every hour? How many people died of asbestosis? How many people are killed in industrial accidents? If you had the same publicity for social conditions that we've had for what's happened to the euro against the pound sterling, the pressure for change would become very, very strong.

Of course, with the worship of money, we get the cult of management. I was told about a boat race between the BBC and a Japanese crew. Both sides practised long and hard to improve their performance and the Japanese won by a mile. So the BBC, being good managers, set up a working party to find out why, and the working party concluded the Japanese had eight people rowing and one steering, and the BBC had eight people steering and one rowing! When managers are faced with a problem of this magnitude there's only one thing to do: appoint consultants. Eighteen months later and after spending a million and a half pounds, the consultants confirmed the diagnosis of the working party, suggested the BBC crew be completely restructured with three assistant steering managers, three deputy steering managers, a director of steering services, and the rower should be given an incentive to row harder! They had another race and the BBC lost by 2 miles. So they laid off the rower for poor performance, sold the boat, and used the money for a higher than average pay award to the director of steering services! Now you can laugh, but I can't think of any organisation that hasn't got this sort of management in place.

Another source of power is power by knowledge, for those who have access to knowledge. That's where the educational system has a very important role to play. I sometimes think that the basis of present educational policy is to train an elite to run the world and train everybody else to take orders. Consider the idea of specially gifted children. I remember in Moscow talking about this to Mr Strokin, the Soviet Minister of Education, because I'd heard they had such special schools in Russia. I asked, 'What about that?' 'Well, as a matter of fact', he said with a funny smile, 'they're actually schools for the children of specially gifted parents'. I thought that was quite a shrewd thing to say!

We also have control of knowledge and information through the media. They explain everything and analyse everything. I was thinking that if Moses had tried to explain the Ten Commandments in an interview on *Newsnight*, he would have never got beyond the second one. If Paxman had said to him:

'What do you think about your mother and father?'

'You should honour your mother and father.'

'Ho! Ho! Honour your mother and father! What about adultery?'

'You shouldn't commit adultery.'

'You don't live in the real world, Moses!'

It is important to be able to hear an argument at length, but the normal interview format makes it almost impossible. Of course even to get on the media there has to be a bit of violence. I was with Jack Jones at a pensioners' rally in Blackpool last May. There were 2000 pensioners (you know, there are 11 million of us). I said to Jack, 'This gathering will not be reported unless you take a brick and throw it through McDonald's, and then there'll be two bishops on *Newsnight* discussing the rising tide of violence among old people!' Of course he didn't throw a brick and there was no coverage. The media in a way encourage violence by only reporting events that produce it.

There is power that comes through science and technology. That is tremendously important, because the development of communicating machines, killing machines, travelling machines and calculating machines has created a huge demand which can't easily be met. There's a technological gap. There's a generational gap; younger people know much more than their parents. The old saying 'Don't teach your grandmother to suck eggs' is absolute nonsense! I depend entirely on my grandchildren to sort out my laptop when it crashes. And the expectation gap. I remember when the Russians put a vehicle on the moon, somebody wrote to me in Bristol and said, 'If the Russians can put a vehicle on the moon, why can't we have a decent bus service in Bristol?' It was a very powerful argument. The question is: who controls the machines? In whose interests are they put? The thing about the killing machines, modern weapons, is that we have a choice in this generation either to deploy our technological ability to meet the needs of humanity, or to destroy the human race. That raises the moral content in all the decisions we take.

Another form of power, in which we begin to see the emergence of some democratic ingredient, is power through organisation. It's quite recent. You've had slavery; you've had feudalism; you've had the Peasants' Revolt; you've had the English Revolution; you've had Tom Paine, whose book *The Rights of Man* was burned by the public executioner because it was seditious; and you've had the Tolpuddle Martyrs. These events have a huge influence on our thinking, though they're not much reported. The Chartists wanted the vote for men and the Suffragettes argued for women. With the development of the Labour movement, you got a new meaning of class, not the old idea of the King and the dukes and the earls and the viscounts and then the common people, but a definition of class in terms of the difference of economic interest between the earners and the owners (the former sometimes called the working class, although nowadays the nature of work has changed so much that if you say 'working class', people think you work in an overall). Nevertheless, if you earn rather than own, you do have a different economic interest, and that was how the Labour Party came to be founded. Party democracy was about the guys at the bottom trying to influence the people at the top, a thing I'll come back to because it didn't make a lot of progress. Similarly, the colonial liberation movements were all about democracy, and self-government, and what a tremendous struggle they had. I've known so many of the colonial national leaders – Gandhi, Nehru, Mugabe, Cheddi Jagan, Nkrumah – all put in British prisons, although they ended up having tea with the Queen. When the demand for self-government first came, it was repressed with considerable force by the colonial power. The demand, both in the old colonial countries and in this country through the development of the Labour movement, was for some democracy, for some control of our future, and for what democracy is about: human equality, representation and accountability.

Somebody asked me recently to define democracy. I thought the best way to put it is like this: if you meet a powerful person – it might be Hitler, Stalin, Bill Gates, anybody you like – ask them five questions. What power have you got? Where did you get it from? In whose interests do you exercise it? To whom are you accountable? How can we get rid of you? That is the democratic question, because if you can't get rid of the people who make the laws you're expected to obey, you don't live in a democratic society.

Consider the way that democracy flowered in Britain, long before the Labour Party was formed in the nineteenth century. Look at the development of local government in, for example, Birmingham after the Municipal Corporations Act of 1837. Democracy in Birmingham flowered in a very interesting way because people used the vote to buy by voting what they couldn't afford individually. They got municipal schools, municipal hospitals, municipal gas, municipal electricity, municipal museums, municipal art galleries, municipal police – that was what it was about. Very rich people have never had to bother with any help from the State to buy a house, or have their children educated, or look after themselves when they're ill or when they're old. But the ballot paper allowed everyone to buy those rights, and that was why it was so important. It got to the point where the Labour movement, having been founded, and then trying to get control of the Parliament by the extension of a franchise, said it wanted a bit of economic democracy as well. That's what the famous Clause 4 was about: 'To secure for the workers by hand or by brain the full fruits of their industry on the basis of the common ownership of the means of production, distribution and exchange.' That didn't really mean nationalisation. I've mentioned that Henry VIII nationalised the Church of England, the Tories nationalised the BBC, the army is nationalised. It's not about nationalisation. It was about some democratic control, i.e. co-operation, municipal enterprise, and so on.

The experience I've had in my life is to find how deeply democracy is distrusted under almost any form of government. I used to go to Moscow as a minister and meet the central committee of the Communist Party. They had not been elected of course. I'd meet the Commissars, and they'd not been elected. I would think 'that's awful'. Then I go to Brussels and meet the Commissioners: they haven't been elected! Meet the Central Bank – they hadn't been elected! Both capitalism and communism allow you to change the management but not to discuss the system. You can only decide whether you want Bush or Clinton, and maybe next Hilary Clinton, or Thatcher, or Major, or Blair or whatever, but you're not really encouraged to discuss any alternative way of running society.

If you look internationally, the developments of the United Nations at the end of the Second World War were an attempt to try and build up some idea of international democratic accountability. The old Concert of Europe was just the colonial powers. The League of Nations were very largely the imperial powers. But when the United Nations was formed, and particularly as the old colonies became liberated, then we did have in the General Assembly something that was beginning to look a bit like the parliament of the world – although of a very primitive kind, and not directly elected because they were sent there by governments. Then came UNESCO, and the importance of saying that war begins in the minds of men and therefore you have to find some way of removing the distrust. Although the UN is very imperfect, it is trying to bring about some change that is likely to be relevant, as with the World Health Organization and Non-Governmental Organisations. We have to look forward to the democratisation of the UN. It seems remote, but remember that the elemental Parliament of 1832 dominated by 2 per cent of the population, all of whom were rich, white men, did in fact gradually get transformed within my lifetime. When I was born, women didn't have the vote until they were 30. Men were so arrogant they said, 'You can't trust the wife until she's 30', so it was only in 1928 that women got the vote at 21. It was only in 1948 that the secondary university vote and the business vote were abolished. We're still not far from the day when we reached one person, one vote. Now we have to take the job on again at a global level.

We'll also have to look again at the question of whether the United Nations should develop an industrial policy. I originally thought sanctions were better than war. They clearly are, but when you see the application of sanctions to a particular country – half a million people thought to have died in Iraq purely because of sanctions – you realise that if sanctions are to be applied for human betterment, they probably have to be applied to multinational corporations and not to nations. If Ford or Gap or Next or these multinational companies couldn't exploit labour in one area in order to undercut the wages in another, then you might make quite a significant contribution. It is a very strange thing that capital can move to wherever the profits are greatest, but people can't move. A company may say to you, 'I'm sorry but wages are lower in Malaysia so we're closing this factory and opening it in Malaysia. I'm afraid that's the law of the market.' But if you live in Malaysia you can't say, 'I understand they're opening a new factory in Cambridge. I want my family to be brought there to get a job and a higher wage.' 'Oh, no, no, no, that's absolutely out of the question!' These matters have to be addressed.

Now look and see how the democratic idea has been eroded. I don't look back on a Golden Age, that would be a foolish thing to do, but I do notice that in the last 50 years there have been quite a number of major assaults on the idea of democracy. The first undoubtedly was the Cold War. We were told the Red Army was planning to come to Cambridge and force everyone here to read Stalin and Lenin instead of the serious academics produced here, and therefore we had to spend billions of pounds on atomic weapons, but I wonder whether that was really true? Was that really what was going to happen? After all, we know now that the Red Army didn't do all that well against Chechnya, so how could they have taken on and defeated NATO? Then I realised that what they were really saying was they didn't want socialist ideas to spread and therefore what they would do would be to present the threat of dangerous ideas as a military challenge, so if you criticised capitalism you were described as an agent of the KGB or working for the Kremlin. This had a profound effect on the future of the world, because the Russians thought we were going to attack them. Indeed one of the reasons why the Soviet Union collapsed was because they spent an enormous percentage of their national income on weapons, thinking they were going to be attacked. It is convenient to have a foreign enemy.

It used to be communism, now it's Islam which helps to strengthen your position at home. It's weird how Islam has been shifted from providing the agents used by the United States to undermine communism, to becoming the main enemy. I remember leading a delegation to protest to the Russian Ambassador, Mr Lunkov, in 1978, at the Russian invasion of Afghanistan, and he said, 'Well, we're only doing it because the Americans have funded people to destroy the regime.' Who were the Americans funding? Osama bin Laden. He was a freedom fighter sent there by the USA in order to destroy communism. If you look back at the relations with the Soviet Union from the very beginning, we sent an army of intervention in after the revolution to destroy the revolution. That's not in our national history curriculum. During the 1930s, the pre-war Conservative government wasn't really appeasing Hitler. If you look at the captured German Foreign Office documents, you will find Lord Halifax, the Foreign Secretary, was sent there in order to congratulate Hitler on his achievement in destroying communism in Germany and standing as a bulwark against communism in Russia. Even in 1941, when the Germans attacked the Soviet Union, President Truman, later the founder of NATO, said in a speech, 'If the Germans seem to be winning, we should support the Russians. If the Russians seem to be winning, we should support the Germans, in the hope that as many as possible kill each other.'

That gives some understanding, looking back on it all, as to what the Cold War was all about. The biggest surprise of all was when I went to Hiroshima, honestly believing, as most people of my generation did, that, awful as the bomb was, it was used in order to save tens of thousands of American lives from the final task of invading Japan. I discovered when I was in Japan that the Japanese had offered to surrender before we dropped the bomb. The only condition they made was that the Emperor should remain. Well, that was what the Americans wanted anyway, to see Japan didn't go communist. But why did they use the bomb? Was it, as I suspect now looking back on it, the first warning-shot in the Cold War, to tell the Russians that the United States had a weapon of supreme power? We must understand the impact of this on our capacity to think clearly about what was really happening.

Similarly, the World Trade Organization, which is now very powerful, does have the power by treaty to dictate to national governments what they can and cannot do in their own societies, and so does the Brussels Commission. If you want to put money into the railways or whatever, you may find you run foul of the rules of the World Trade Organization that says you can't subsidise your own industries. These are erosions of democracy on an absolutely massive scale. I remember way back in 1975, when I was a minister and produced a White Paper on industrial policy, I was told by the Foreign Office that I couldn't publish a bill about industrial policy until it had been cleared with the Commission in Brussels. When Parliament can't even know what a minister wants to do until it's been cleared by people who've never been elected, you are beginning to see a threat to the survival of democracy.

I was on the Council of Ministers for four years, indeed I was the President of the European Council of Energy Ministers, and I found it an absolutely fascinating experience, for two reasons. First of all, it was the only committee I ever sat on where I was not allowed to put in a document. Only the Commission could put in a paper and, as a minister, you could only say 'yes' or 'no' to what they offered. All the laws are made, in secret, by the Council of Ministers. During my presidency I wrote to every other minister and said, 'Let's meet in public'. They nearly had a nervous breakdown, because the idea that the press and the public would know what was going on in the Council of Ministers terrified them! And yet the laws in Brussels are not made by the European Parliament, but made by the ministers. It's the only parliament in the world that meets in secret, and the directives and decisions of the Council of Ministers can repeal laws we've passed in our Parliament and every other parliament and can impose laws on us that have never been discussed by our Parliament. That is, in my opinion, a very significant erosion, never discussed in this way.

Of course when the euro comes in, and we have the Central Bank controlling everything, the Chancellor of the Exchequer will simply become the chairman of the finance committee of a sort of local authority based in London. That's why Gordon Brown may not be very keen on it. Similarly, the Governor of the Bank of England is not very keen on it. If you want to see the possible consequences of that, you've only got to look at the Argentine, where the economic crisis they have to cope with came because the peso was linked to the dollar, and whereas Brazil was free to devalue and boomed, the Argentinean economy collapsed. There were 5 presidents in 12 days, which gave you some indication about the extent to which some democratic input into it was frustrated.

Quite apart from all that, there is the centralised executive power now found in Britain. The present government follows a tradition established by Mrs Thatcher. Peter Mandelson, when he was a minister just after the election in 1997, made a speech in Germany in which he said, 'The era of representative democracy is coming slowly to an end.' It was a very important statement, because what he was saying was that legitimacy doesn't any longer depend on being elected, it depends on being efficient. He was citing the Central Bank. And Parliament is now in decline. I left Parliament, I have said, to devote time to politics. I thought at the time I would miss the chamber of the House of Commons, but I don't. The Speaker has very kindly said, as I've been there 50 years, I can use the building. I go to the tea room and the library and meet all my mates. This is the normal right of every peer who has been an MP, so I have all the benefits of peerage without the humiliation of being a Lord!

When I look back on the chamber, I don't miss it at all. I've no urge to ask clever supplementary questions of the Prime Minister. Indeed, Prime Minister's Questions switch me off completely. 'Could I trouble my right honourable friend to remind me of his latest triumph since last Wednesday?' They say there's so many planted questions, it ought to be renamed Gardener's Question Time! And people are beginning to notice. They don't think it matters any more. The Cabinet is so weak. The Cabinet meets for half an hour or 20 minutes now, to be told what the Prime Minister has decided – if he hasn't got a photo opportunity that makes it difficult for him to be there. I looked in my diary: in January 1968, we had eight full-day meetings of the Cabinet in a single month, morning and afternoon. You may say it took a lot of time, but the people round the table, Dick Crossman, Roy Jenkins, Gerald Gardiner, Barbara Castle – each formidable people, worth listening to, but the key to success was that the decisions that were made were collective decisions. Even if you were defeated, and you were very often defeated in the Cabinet, you were committed to the outcome because you'd been able to participate. All that has gone, and is replaced by patronage.

Prime Ministerial patronage. I mentioned the appointment of the Archbishop of Canterbury. That goes back to the nationalisation of the Church of England, but the Prime Minister has put 248 people into Parliament. It takes the whole population of the country to elect 650 MPs and it's taken one Prime Minister to put 248 people into the House of Lords. Now, whatever you may say about these people as peers – they may be very, very interesting, and I'm not saying a word against any of them – it is nothing whatever to do with democracy. It is a return to a medieval system, because, although not many people realise this, when peerages began they were not hereditary. The king made somebody a peer for life and then put another one in. The hereditary peerage came in later. So we are now modernising ourselves back to the fourteenth century, namely that the King, or the Prime Minister now, stuffs the House of Lords with his friends. Of course that's a very convenient thing to do because if you've got a newspaper proprietor who's being a bit difficult, a peerage could help – that's why they call them Press Lords. If you want a vacancy to be created in the House of Commons, you offer the sitting member a peerage and then you parachute Sean Woodward in or whoever it happens to be. You realise how convenient patronage is, and also realise it has nothing whatever to do with the democratic process.

In the party the centralisation of power is going on on a huge scale. There is somebody, Charles Clark, who's been announced as Chairman of the Labour Party. He's never been elected, but he's been called Chairman and he is really the Prime Minister's boss of the Labour Party. I think this is what people feel. I certainly felt it very strongly when I was there and so did my constituents. Instead of being represented, we are being managed, and there's all the difference in the world between electing someone to represent you and electing someone who's going to manage you. I used to get a fax every day from the Millbank Tower when I was a Member. They took a lot of trouble; they drafted it with a quote from me. You know, 'Tony Benn welcomes compulsory homework for pensioners', or whatever the latest gimmick was, and I was expected to take it out of the fax machine, and fax it back again to the *Derbyshire Times* in the hope they'd print it. I did make a speech once saying, 'I feel less and less like a Member of Parliament, and more and more like an Avon Lady who's told what to say when she knocks at the door'! I had a furious letter from the Managing Director of Avon, bitterly complaining that I had compared Avon Ladies to Members of Parliament!

Now we come to the growth of the American empire. The American empire is the greatest empire the world has ever seen. Most empires attempt to get hold of resources. I did a debate at the time of the war in Kuwait with Enoch Powell, who was a classical scholar, about the Peloponnesian Wars, of which I knew absolutely nothing. But he explained that at the time of the Peloponnesian Wars wood was essential for warships and therefore people had wars for wood. And of course if you look at recent wars they've almost all been about oil. As Energy Secretary I was called to the Cabinet committees on the Falklands. I said, 'What have I to do with the Falklands?' 'Ooh', said my Permanent Secretary, 'there's more oil round the Falklands than there is around the United Kingdom'. The Falklands War was a war for oil. The Gulf War was a war for oil. The Afghan War was a war for oil because, three or four years ago, when Bush was Governor of Texas, the Texans invited the Taliban to come to discuss a pipeline from the Caspian oil to the Western market.

That's what it's about. The UN has been bypassed and civil liberties eroded. I don't want to go in any great detail to these prisoners in Guantanamo, but if you have a war against terrorism you'd think if you caught somebody he was a prisoner of war. It turns out now they're a new category called 'Illegal Combatants'. The erosion of civil liberties is very, very frightening and I think what we are now up against is revenge posing at justice. There is a danger in this global alliance against terrorism, which is perfectly understandable, but all the top people in the world are huddling together against any form of dissent in their own countries. I'm sure in Beijing, they'd say, 'That's a good idea. We'll join in the battle against terrorism, and that will help us against the Falun Gong'. You can see, it wouldn't take an awful lot of time for people to begin to realise that you could define terrorism in almost any way you liked. Eco-terrorism is a word I've heard. Maybe the people who go to Seattle or Genoa are thought of as terrorists. But if you criminalise dissent then you are in a very dangerous situation. Historically, before the democratic impetus began to be felt, the argument was between the government of the day and the people, and now I think that argument is reappearing. It is the governments of the world against the peoples of the world, and this is done in order to maintain the status quo, whatever is convenient to that particular government. Once you begin defining opposition as terrorism, then you create problems.

I remember speaking as a young MP in 1964 in Trafalgar Square, supporting a very well-known terrorist who'd just been convicted of the crime of terrorism, to which he confessed. I was duly denounced in the *Daily Mail* and all that. The next time I met him he had a Nobel Peace Prize and was President of South Africa. When Gandhi was in London in 1931, I met him, when I was six. A journalist said to Gandhi, who had just come out of a British prison and was on his way back in again, 'What do you think of British civilisation, Mr Gandhi?' And Gandhi said, 'I think it would be a very good idea'! Freedom fighters are people who are driven to use force in order to get democracy, just as the imperial powers use force to eliminate democracy, and this is something that has to be thought about very carefully.

The last points I want to make are about the possible impact of a world recession on democracy. I was brought up in a very political family. I bought *Mein Kampf* when I was about 10 and I've still got it in the shelf at home. I've got Mussolini's biography, with a foreword by the American Ambassador in Rome saying Mussolini was the greatest figure of our time and sphere. After the effect of the war on Italy in the '20s and in Germany in the '30s, 6 million unemployed, it was very easy for somebody to come along and find a scapegoat – the Jews, the communists, whatever – and say to the Germans as Hitler did, 'I'll give you work.' And he did. He gave them work as soldiers – which cost millions of lives in the war.

Hitler said, 'Democracy inevitably leads to Marxism.' Now you work that one out! If you think what happened in Stalin's Russia, what he was saying was that if you give people power, they'll move to the left, and that's why democracy is so controversial. I think that the anti-globalisation movement, in a very simple and crude way, is a movement for global democracy.

These people have no common ideological position. The unions, the churches, the peace movement, the environmental movement, and so on – they come together because, they say, the exercise of power in our world is done without any regard to our interests, and indeed without our consent, and what they're trying to do is to put pressure on governments.

I don't believe that the answer to these problems is a tight ideological one. I'm a Socialist. But I went to a meeting not long ago, a Socialist conference in London, about 7000, a huge meeting, and somebody said, 'We're all victims of capitalism. We've got to smash the State.' I said, 'That's a very interesting proposition, but if an old lady comes to me in Chesterfield and says, "Tony, I'm 80 now, my husband just died. Could I have a bungalow?" and I say to her, "Well, you're a victim of capitalism. We must smash the State", she would say, "Tony, that's very, very interesting, but what are the prospects of a bungalow?"' You actually have to approach people through their own experience, not through trying to impose an ideological solution on everybody's problem. That I think is what is needed.

My own experience of progress over the years is that if you come up with a progressive idea, to begin with it's absolutely ignored. The media don't mention it, and if you go on you're mad, absolutely bonkers. I've been accused of that myself. If you go on after that, you're very dangerous. Then there's a pause, and then you can't find anyone who doesn't claim to have thought of it in the first place! And that is how progress occurs, it begins from the bottom. I think of the environmental movement – it's a very vivid example. If you take someone like Swampy, for example, 10 years ago he was a bearded weirdie who'd be taken into custody by the local constabulary. He'll be in the House of Lords next because the argument has been won. No government now dare ignore the environment, even if Bush won't go along with the Kyoto Agreement. My experience of Parliament is that it's the last place to get the message. That's why I'm working at another level now, because when Parliament decides something you can be sure that five years ago everybody else had reached the same conclusion.

We have to try and accelerate the process of justice from the bottom to the top. There are opportunities now, the Internet is very significant in undercutting the Murdochs and the CNNs and has great potential. Certainly I find things on it which I would never be able to discover in any other way. Communication and access to knowledge is so important. But it will be a very long and hard struggle. I don't think that people with power ever, ever, want to give it up, and you have to persuade them and put such pressure on them that they feel their own survival requires concessions. Democracy to me is a means, not an end. It's a route map and not a destination. It's a journey. It's a way by which you discuss things and how you decide them. For that, you have to have understanding. You have to have self-organisation because you won't get any help from the people at the top – you've got to do it yourself. That's the lesson of all the great changes – anti-colonial movement, women's movement, environmental movement, trade union movement – they've always begun at the bottom and got through to the top.

It is very hard. There are moments when I write my diary at night, and I put such gloomy things in it that I have determined they shall never be published. On the other hand, there are moments when it's so exciting that I wonder if it's good for someone of my age to get as excited as I do. Hope is the fuel of progress, and fear is a prison in which we put ourselves. If we approach the problems of political progress in that spirit, then I think we shall do quite well.

Notes on the contributors

Mary Archer read Chemistry at Oxford and then Imperial College, London, and taught the subject for 10 years at Newnham and Trinity Colleges, Cambridge. She is a Visiting Professor at the Imperial College Centre for Energy Policy and Technology and a member of the Royal Society's Energy Policy Working Group. She is also President of the National Energy Foundation and of the UK Solar Energy Society, and the recipient of the 2002 Melchett Medal of the Institute of Energy.

Tony Benn, born into an aristocratic family and educated at Westminster and New College, Oxford, trained as a pilot during the Second World War. He was elected to Parliament in 1950. Anticipating that inheritance of his father's peerage would disqualify him from continuing to serve in the House of Commons, he campaigned for a bill to permit him to renounce the title; this struggle lasted until 1963 when the Peerage Act was passed. As MP for Bristol South-East, he served as a Cabinet minister in the Wilson and Callaghan governments from 1964 to 1970 and from 1974 to 1979. In 1984 he became Member of Parliament for Chesterfield, and increasingly took on the role of unofficial leader of the Labour Party's radical left. In 2001 he gave up his parliamentary seat because he 'wanted to devote more time to politics'. His diaries and essays have been published in volumes including *Free Radical* (Continuum International Publishing, 2003), *Tony Benn Diaries* (six volumes, abridged, Hutchinson, in 1996) and *Free at Last* (Hutchinson, 2002), in addition to a biography by David Powell, published in 2001 by Continuum International.

Elisabeth Bronfen is a graduate of the University of Munich and currently holds the Chair of English and American Studies at the University of Zurich. She has been a Guest Professor at Columbia University, Princeton University, Sheffield

Hallam University, the University of Copenhagen and the University of Aarhus. While a specialist in nineteenth- and twentieth-century literature, her articles range from work on gender studies, psychoanalysis and film, to cultural theory and art. Her publications include *Over Her Dead Body: Death, Femininity and the Aesthetic* (Manchester University Press, 1992) and the collection of essays entitled *Death and Representation*, coedited with Sarah W. Goodwin (Johns Hopkins University Press, 1993). Professor Bronfen has also published books on hysteria, Sylvia Plath and Dorothy Richardson; she has edited a four-volume German edition of the poetry and letters of Anne Sexton and coedited a collection of essays on recent scholarship in gender studies (*Feminist Consequences: Theory for the New Century* (Columbia University Press, 2000)). Her most recent book is *Home in Hollywood: The Imaginary Geography of Cinema* (Columbia University Press, 2004).

John Conway is von Neumann Professor of Mathematics at Princeton University. Educated at Cambridge University, he served on the Mathematics faculty there prior to joining Princeton in 1986.

He is one of the pre-eminent theorists in the study of finite groups (the mathematical abstraction of symmetry) and one of the world's foremost knot theorists. In his many books and journal articles he has broken new ground in number theory, game theory, coding theory, tiling and the creation of new number systems. The system of 'Surreal Numbers' which he invented is the subject of a popular book by computer scientist Donald Knuth.

Beyond the academic world, Conway is widely known as the inventor of the 'Game of Life'. He may well have the distinction of having more books, articles and Web pages devoted to his creations than any other living mathematician.

He is a Fellow of the Royal Society and received the Polya Prize of the London Mathematical Society and the 1998 Frederic Esser Nemmers Prize in Mathematics.

Neil deGrasse Tyson is the Frederick P. Rose Director of the Hayden Planetarium in New York and a Visiting Professor in the Department of Astrophysical Sciences at Princeton University. He received his first degree in Physics from Harvard University, his MA in Astronomy from the University of Texas, and his Ph.D. in Astrophysics from Columbia University. In 2004, Tyson was appointed by President Bush to serve on a nine-member commission on the Implementation of the United States Space Exploration Policy, dubbed the 'Moon, Mars, and Beyond' commission. This group navigated a path by which the new space vision can become a successful part of the American agenda.

He is especially well known for his monthly column in *Natural History* magazine and his popular science books, the first of which, *Merlin's Tour of the Universe* (Columbia University Press, 1989), has been translated into Chinese, Italian, Polish, German, Spanish and Japanese. His most recent books are the memoir *The Sky Is Not the Limit: Adventures of an Urban Astrophysicist* (Doubleday, 2000), *Cosmic Horizons: Astronomy at the Cutting Edge* (New Press, 2001), coauthored with Steven Soter, and *One Universe: At Home in the Cosmos* (Joseph Henry Press, 2000), coauthored with Charles Liu and Robert Irion. Tyson's contributions to the public appreciation of the cosmos have recently been recognised by the International Astronomical Union in their official naming of asteroid '13123 Tyson'.

Derek B. Scott is Professor of Music at Salford University. He has worked professionally in radio, TV, concert hall and theatre. He is the author of the books *From the Erotic to the Demonic* (Oxford University Press, 2003) and *The Singing Bourgeois* (2nd edn, Ashgate, 2001), editor of *Music, Culture and Society: A Reader* (Oxford University Press, 2000), and author of numerous articles concerning music and ideology. He has been at the forefront in identifying changes of critical perspective in music and sociology. He is the General Editor of Ashgate's Pop and Folk Series, and a member of the Editorial Boards of *Popular Musicology* and the Internet *Critical Musicology* Journal. He was a founder member of the UK Critical Musicology Group in 1993 and organiser of their first major conference in Salford in 1995. He is also a composer, whose works include two symphonies for brass band.

Maureen Thomas is Creative Director of the Cambridge University Moving Image Studio (CUMIS); Senior Creative Research Fellow at the Narrativity Studio, Interactive Institute, Sweden; and Associate Professor in Interactivity and Narrativity at the Norwegian Film School. From 1993 to 1998 she was Head of Screen Studies at the National Film & Television School, UK.

A dramatist and director for stage, radio and film, in 2000/1 Maureen made a feature-length interactive digital video hypermovie, *Vala*, which involves interactors directly, theatrically and actively in a narrative game of chance, destiny and time-travel. (First demo prototype, Nedslag, Fylkingen Centre for New Music and InterMedia Art, Stockholm (May 2000); full-length demo prototype, Electrohype, Malmö, Sweden, and Rencontres Electroniques, Rennes, France (October 2000); full interactive prototype Arts Picturehouse Cinema Cambridge International Film Festival UK (July 2001), and Nordic Interactive Expo, Copenhagen, Denmark (October 2001).)

Maureen's analysis of the relationship between cinematic narrative and the structures of 35 interactive adventure-games has been published in the book *Architectures of Illusion* (Intellect Books, 2002), which provides a profile of the work of CUMIS.

Index